Unmanned Aerial Systems in Agriculture

Unmanned Aerial Systems in Agriculture
Eyes Above Fields

Edited by

DIONYSIS BOCHTIS
Institute for Bio-Economy and Agri-Technology (IBO),
Centre for Research and Technology Hellas (CERTH),
Thessaloniki, Greece

ARISTOTELIS C. TAGARAKIS
Institute for Bio-Economy and Agri-Technology (IBO),
Centre for Research and Technology Hellas (CERTH),
Thessaloniki, Greece

DIMITRIOS KATERIS
Institute for Bio-Economy and Agri-Technology (IBO),
Centre for Research and Technology Hellas (CERTH),
Thessaloniki, Greece

ELSEVIER

ACADEMIC PRESS
An imprint of Elsevier

ISBN: 978-0-323-91940-1

For Information on all Academic Press publications
visit our website at https://www.elsevier.com/books-and-journals

Publisher: Nikki P. Levy
Acquisitions Editor: Nancy J. Maragioglio
Editorial Project Manager: Catherine Costello
Production Project Manager: Rashmi Manoharan
Cover Designer: Mark Rogers

Typeset by MPS Limited, Chennai, India

Working together
to grow libraries in
developing countries

www.elsevier.com • www.bookaid.org

Contents

Section C Operational aspects

Section D Sustainability aspects

List of contributors

Lefteris Benos
Institute for Bio-Economy and Agri-Technology (IBO), Centre for Research and Technology Hellas (CERTH), Thessaloniki, Greece

Dionysis Bochtis
Institute for Bio-Economy and Agri-Technology (IBO), Centre for Research and Technology Hellas (CERTH), Thessaloniki, Greece

Chris Cavalaris
Department of Agriculture Crop Production and Rural Development, University of Thessaly, Volos, Greece

Nikoleta Darra
Department of Natural Resources Management and Agricultural Engineering, Agricultural University of Athens, Athens, Greece

Spyros Fountas
Department of Natural Resources Management and Agricultural Engineering, Agricultural University of Athens, Athens, Greece

Željana Grbović
BioSense Institute, University of Novi Sad, Novi Sad, Serbia

Lizhen Huang
Department of Manufacturing and Civil Engineering, Norwegian University of Science and Technology, Gjøvik, Norway

Bojana Ivošević
BioSense Institute, University of Novi Sad, Novi Sad, Serbia

Vassilis G. Kaburlasos
Human-Machines Interaction (HUMAIN) Lab, Department of Computer Science, International Hellenic University (IHU), Kavala, Greece

Aikaterini Kasimati
Department of Natural Resources Management and Agricultural Engineering, Agricultural University of Athens, Athens, Greece

Dimitrios Kateris
Institute for Bio-Economy and Agri-Technology (IBO), Centre for Research and Technology Hellas (CERTH), Thessaloniki, Greece

Marko Kostić
Faculty of Agriculture, University of Novi Sad, Novi Sad, Serbia

Michael Gerasimos Koutsiaras
Department of Natural Resources Management and Agricultural Engineering, Agricultural University of Athens, Athens, Greece

George Kyriakarakos
farmB Digital Agriculture S.A., Thessaloniki, Greece

Pekka Leviäkangas
Civil Engineering, Faculty of Technology, University of Oulu, Oulu, Finland

Nataša Ljubičić
BioSense Institute, University of Novi Sad, Novi Sad, Serbia

Ari Lomis
Department of Natural Resources Management and Agricultural Engineering, Agricultural University of Athens, Athens, Greece

Marko Panić
BioSense Institute, University of Novi Sad, Novi Sad, Serbia

George Papadopoulos
Department of Natural Resources Management and Agricultural Engineering, Agricultural University of Athens, Athens, Greece

Vasilis Psiroukis
Department of Natural Resources Management and Agricultural Engineering, Agricultural University of Athens, Athens, Greece

Anssi Rauhala
Civil Engineering, Faculty of Technology, University of Oulu, Oulu, Finland

Georgios Siavalas
Human-Machines Interaction (HUMAIN) Lab, Department of Computer Science, International Hellenic University (IHU), Kavala, Greece

Aristotelis C. Tagarakis
Institute for Bio-Economy and Agri-Technology (IBO), Centre for Research and Technology Hellas (CERTH), Thessaloniki, Greece

Anne Tuomela
Civil Engineering, Faculty of Technology, University of Oulu, Oulu, Finland

G. Vasileiadis
farmB Digital Agriculture S.A., Thessaloniki, Greece; Department of Hydraulics, Soil Science and Agricultural Engineering, School of Agriculture, Aristotle University of Thessaloniki, Thessaloniki, Greece

Eleni Vrochidou
Human-Machines Interaction (HUMAIN) Lab, Department of Computer Science, International Hellenic University (IHU), Kavala, Greece

Wendy Wuyts
Department of Manufacturing and Civil Engineering, Norwegian University of Science and Technology, Gjøvik, Norway

Preface

Agricultural production has rapidly changed in the last decades following the need for intensification to meet the increasing food and agri-products demand. The evolution of agri-technologies, within the context of improving the efficiency of agricultural production, encourages the adoption of new methodologies and tools for the performance of agricultural tasks. Intelligent machines, autonomous vehicles, innovative sensing, and actuating technologies, along with improved information and communication technologies, have led to a complete restructuring of agricultural production with decision-making tools and practices now taking a prominent role in the agri-business. Toward that direction, agricultural production processes, including precision applications based on mapping of soil and plant attributes, are becoming popular.

Although advanced features of precision farming have been adequately described and analyzed in various scientific reports, the issue of unmanned aerial systems (UAS) for agricultural applications needs to be viewed in a different, more spherical, and integrated way. In this light, this book aims at bridging the gap on the disseminated knowledge in the use and operations of UASs in agri-production during the last decades, as well as their short-term future applications. The book approaches aspects of digital farming facilitated using autonomous aerial platforms in a holistic way, contributing to the development of integrated agricultural systems, using information acquired by aerial remote sensors. It starts with a general overview of the unmanned aerial vehicles (UAVs) in agriculture and continues with a deep look into the world of UAS including energy, economical, and social aspects from their use in agri-production systems. By covering theoretical and technical aspects, as well as including case studies and real-world insights, the book provides the opportunity for the reader to develop a clear understanding of how agriculture production can meet sustainability goals in the modern business world, while assessing potential opportunities from the exploitation of available modern technologies such as UAS in agri-production. The book is organized into four main thematic areas (sections), namely, Introduction, Applications in Agriculture, Operational Aspects, and Sustainability Aspects.

In the first section, a general overview of the utilization of UAS in the agricultural sector is provided. Chapter 1 introduces the readers to the drones' evolution by providing an insight into the technological and applicational aspects from their first existence to their wide range of applications they support nowadays with focus in agricultural applications. Subsequently, a classification is presented based on the weight and size, aerodynamic features, and application. Finally, the limitations of the existing drones and challenges that need to be addressed are discussed from a broader perspective.

Chapter 2 introduces the types of UASs and their potential onboard carried sensors. It also provides an insight into operational aspects such as mission planning and discusses the state-of-the-art technology used in UASs with their application in the field of precision agriculture, and future perspectives. UASs are equipped with various sensors. These sensors allow for the collection of detailed and high-resolution data on crop health, soil moisture, and crop yields. Overall, UASs equipped with sensors are a valuable tool in precision agriculture, allowing for more efficient and effective crop management. Therefore a discussion on the UAS-enabled sensing systems is provided.

The second section provides an overview of the practical applications of UAV-enabled systems in the agricultural sector. It starts with Chapter 3, which discusses the use and potential applications of the UAS in precision agriculture. The applications may include crop growth monitoring, disease and weed detection, irrigation or fertilization, plant counting, end-of-season yield and biomass estimation, and zone management delineation, among several others. Knowledge and understanding about the usage and limitations of UAS data are necessary as the corresponding market grows in a fast manner. Farmers' adoption and future perspective of UAS in precision agriculture are also discussed.

The following chapters of this section refer to real applications of UAS in specific crops. Chapter 4 provides a deeper view into the use of such aerial systems in specific applications in horticultural and arable sectors, by initially describing fundamental principles on surveying aspects for these applications. An extended overview of specific use cases and active research are also included. Overall, this chapter aims to assist both agricultural researchers that want to incorporate UAV methods in their experimental approaches, as well as individual farmers interested in the capabilities of this new technology.

Chapter 5 focuses on the use of UASs in precision agriculture, particularly for fruit tree growth management. The use of UAVs in agriculture is becoming increasingly popular, especially for applications in tree crops as it can provide detailed and accurate information on fruit crop health and resource use, while reducing labor and increasing yields. The chapter also looks at the use of drones for pest control and irrigation efficiency, as well as for yield forecasting and selective harvesting. The importance of flight planning is also discussed.

The third section of the book is focused on "Operational Aspects." In Chapter 6, a review of some of the most popular applications of UASs in precision agriculture during the last 5 years is provided. More specifically, single and multiple UAS applications are reviewed in both monitoring applications and spraying applications. The main characteristics of UASs are presented; data acquisition methods, supportive algorithms, and technologies are also summarized. The objective of this work is to investigate the potential of UAV systems for a cost-effective implementation of precision agriculture practices while, furthermore, to discuss their limitations as well as a series of future trends.

Chapter 7 elaborates on the usage of UASs for multiple purposes in the sectors of open-field agriculture, forestry, and livestock farming. It provides the concept for the development of an intelligent ecosystem to address the multiple purposes feature in the light of deploying UAVs to promote sustainable digital services for the benefit of a large scope of end users.

Digital twins are a recently developed technological concept that can be used for systems analysis and decision support. Therefore Chapter 8 overviews several recent applications of using digital twin coupled with UAVs in agriculture. Overall, digital twins are still in their infancy in the agricultural context, mainly because of the complex and dynamic nature of the sector itself combined with the slow adoption of digital technologies. It is anticipated that open-source digital twins and information of think platforms are going to foster scientific and technological progress in this field.

The last section of the book is devoted to "Sustainability Aspects" of the UAS in agriculture. The section opens with the "challenges and opportunities for cost-effective use of UAS in agriculture" (see Chapter 9) which elaborates on the cost of using UASs and the expected economic benefits. This is the most crucial aspect manifesting the widescale adoption of UASs in agriculture. In this chapter, the analysis of the main cost components of utilizing a UAS facility in agriculture is provided. The analysis addresses the direct variable and fixed cost associated with, for example, ownership and labor energy, among others, as also indirect costs related to transportation, data handling and analyzing, and others. It continues by discussing the limitations and challenges for achieving efficient and cost-effective field UAS operations, also making comparisons to the available alternatives, such as satellite observations, for the case of monitoring operations, and conventional machinery, in the case of material applications. The chapter ends by analyzing the current state and future perspectives of UASs and how they are expected to alter the agricultural sector in the short-term future.

Apart from the cost-related aspects, flight autonomy is considered one of the main constraints in the utilization of UASs in all domains and in agriculture as well. Energy efficiency is one of the greater challenges for the UASs to become efficient and economically profitable. Chapter 10 presents the main architecture of different kinds of UASs and analyzes their power and energy demands and their suitability for alternative tasks. Focusing on different types of propulsion systems, it analyzes their performances and response to power demand variations. The chapter also presents different types of energy storage components and the technologies employed for providing electric power in UASs, their capacity, and main features and discusses the key factors affecting drone power demands and energy consumption. Finally, the most promising future technologies, implemented to improve drone endurance and additional to provide a clean energy source, are highlighted presenting also some commercially available solutions.

Gender-related issues in the development and utilization of technological developments in agriculture and other domains are a requirement in the context of raising awareness. Funding schemes, such as the ones by the European Commission, are steering toward integrating a gender dimension in research, development, and innovation projects. Chapter 11 provides an overview of definitions, operational rules, and recommendations that are relevant to gender perspective and tailored to drone and digital twin engineers, operators, and researchers in smart forestry and agriculture. Furthermore, apart from the gender dimension, there are evident threats and risks related to the use of drones posed to, for example, physical safety, privacy, and data security. This has made enhancing UAS governance a necessity.

The final chapter, Chapter 12, provides a detailed overview of the regulatory framework in the European Union on common rules in the field of civil aviation, which applies to all UAS. Furthermore, it discusses the challenges and social acceptability of drones and the regulations concerning the risks associated with data protection and safety, and the effect on the wider use of such systems.

It is a common conviction that autonomous aerial systems will play a leading role in the course of digitizing agriculture toward the future of farming within the Agriculture 4.0 concept. This book reflects the joint efforts of all authors here to cover the major aspects and concerns related to the use of UASs in agriculture, providing the most recent trends and developments of the associated technologies and their related applications, covering both sustainability and societal perspectives.

Dionysis Bochtis
Aristotelis C. Tagarakis
Dimitrios Kateris

Introduction

CHAPTER 1

Developments in the era of unmanned aerial systems

Lefteris Benos, Dimitrios Kateris, Aristotelis C. Tagarakis and Dionysis Bochtis
Institute for Bio-Economy and Agri-Technology (IBO), Centre for Research and Technology Hellas (CERTH), Thessaloniki, Greece

Introduction

An unmanned aerial vehicle (UAV), commonly known as drone, can be simply defined as an unmanned flying vehicle, which can be remotely operated, fly via preprogrammed flight plans, or can operate autonomously under a mission-oriented plan. Taking information from sensors (like accelerometers and gyroscopes) and the global positioning system (GPS), the flight controller comprehends its orientation and position. The rate of acceleration is defined using accelerometers, while gyroscopes serve to keep the drone flying or hovering smoothly. Most drones can be controlled using smartphone applications and a controller used to control them and keep track of the flight telemetry. At the same time, drones can exploit a variety of sensors and cameras, hence, supplying them with remarkable observation-making capabilities to be utilized for a wide range of operations. Software algorithms can convert the images into three-dimensional (3D) maps allowing for collision avoidance. A broad spectrum of sensors is utilized including vision, infrared, ultrasonic, LIDAR (an acronym for "light detection and ranging"), and time of flight sensors. The data acquired by the sensors are intelligently combined to optimize the performance of the system.

Currently, there is no official classification of drones, although some suggestions have been presented in the scholar literature based on aspects like weight, size, flight range, and aerodynamic features (Brooke-Holland, 2012; Hassanalian and Abdelkefi, 2017a; Watts et al., 2012; Arjomandi et al., 2006). The use cases of these vehicles seem endless, since they can be indicatively used for:

1. Industrial areas inspections (Nooralishahi et al., 2021; Li et al., 2021).
2. Security surveillance (Reyes et al., 2017; Lo et al., 2021).

Unmanned Aerial Systems in Agriculture
DOI: https://doi.org/10.1016/B978-0-323-91940-1.00001-3

3

3. Emergency response (Khan et al., 2021; Ganesh et al., 2021).
4. Traffic monitoring (Huang et al., 2021; Beg et al., 2021).
5. Telecommunications (Naqvi et al., 2018; Fotouhi et al., 2019).
6. Atmospheric research (Lambey and Prasad, 2021; Chechin et al., 2021).
7. Agricultural applications (Kaivosoja, 2022; Radoglou-Grammatikis et al., 2020).
8. Forestry remote sensing (Guimarães et al., 2020; Cromwell et al., 2021).
9. Military usage (Sharkey, 2011; Konert and Balcerzak, 2021).
10. Small items' delivery (Saeed et al., 2021; Zhang et al., 2021).
11. Professional aerial cinematography (Gschwindt et al., 2019; Goh et al., 2021).
12. Sports broadcast (Wang et al., 2017; Zachariadis et al., 2017).

Seeing it from a value-added system perspective, drones can constitute an integral part of the so-called Internet of things (IoT) (Labib et al., 2021). In this case, these so-called unmanned aerial systems promise an alternative solution to the restrictions of IoT terrestrial infrastructure and, thus, new services and opportunities for several market sectors and society (Anush Lakshman and Ebenezer, 2021). Among the main reasons leading to the radical development of drones is the considerable advancement of sensory technologies and wireless sensor networks as well as the improvement in battery performance and the drop in price. Features such as weight, wingspan, loading, endurance, and maximum altitude and speed are design parameters of central importance (Hassanalian and Abdelkefi, 2017a). Finally, the beneficial realization of drones' potential does not depend on a sole technology, but on a plethora of interconnected procedures and systems in accordance with innovative approaches and international regulations.

This chapter aims at providing insight to the evolution of drones, the classification of them based on several aspects according to the relative literature and the main challenges associated with drones. The remainder of this chapter is structured as follows. Section "History of drones" briefly describes the history of drones from their forerunners and initially military-purposed aerial vehicles up to the commercial drones. Section "Classification of drones" focuses on the categorization of drones based on weight and size, aerodynamic features, and application. Finally, the chapter ends with a section, where concluding remarks are drawn along with a discussion on the existing challenges and opportunities.

History of drones
Forerunners of drones

As can be seen in Fig. 1.1, drones' history goes back to 1483 and, in particular, to the so-called air gyroscope designed by Leonardo Da Vinci. As he said: "I have discovered that a screw-shaped device such as this, if it is well-made from starched linen, will rise in the air if turned quickly." In fact, this invention is regarded to be the precursor of modern multicopters flying through buoyancy taking place at the ends of the rotor blades.

Formally, the drones' chronicle starts in 1783 in France, where Montgolfier brothers made a public demonstration flight of a globe-shaped balloon filled with smoke. These balloons, which were later named as airships, played a key role in the history of aeronautics. In 1849 Austrian soldiers attacked Venice by using unmanned balloons filled with hot air and equipped with explosives. The forerunner of modern remote-controlled drones, however, is considered as the first radio-operated boat; a technological masterpiece showed by Nikola Tesla in Madison Square Garden in 1898. Tesla came to this invention while examining radio waves. More specifically, he designed a wave generator, which was called a Tesla coil. To exhibit its usage, he firstly used it to wirelessly illuminate lamps. Tesla realized that if he could power lamps wirelessly, he could also turn other devices on and off and, consequently, control them.

Drones for military operations

The earliest unmanned radio-controlled aircraft, which appeared in World War I, was Curtiss N-9 invented by Cooper and Sperry, while

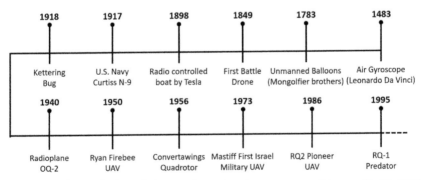

Figure 1.1 The evolution of drones; from the air gyroscope of Leonardo Da Vinci and unmanned balloons to military drones.

Kettering conceived the so-called Kettering Bug that is seen as a predecessor of cruise missile. Few years later, in the context of World War II, the Radioplane OQ-2 was developed, the first mass-produced drone in the United States. A subsequent version of OQ-2 created by the retired aircrafts, namely the OQ-3, was converted into aerial vehicles aiming at reconnaissance missions in 1945. After the close of World War II, the United States produced a series of target drones, known as Ryan Firebee, which were the first jet-propelled drones. These aerial vehicles were used for air-to-air combat training. In the mid-1950s, the first four-rotor helicopter, known as Convertawings Quadrotor, was presented whose four rotors had an "H" position.

A pioneer UAV was also the Mastiff first Israeli Military UAV being equipped with live-streaming capabilities, which was very challenging at that time (1973). Initially, a black-and-white video was able to be transmitted, but later, a colored camera was added. As a consequence, a new role had been assigned to drones toward tracking, for instance, a human or a vehicle. In 1986 the RQ2 Pioneer UAV was used by the US Navy to provide real-time images of the battle, by also carrying out several tasks including reconnaissance and surveillance of targets. After the 1990s, as countries had recognized the significance of UAVs in wars, they utilized lightweight materials in conjunction with advanced technology regarding communication and signal processing. Indicatively, the famous RQ-1 Predator was developed by the United States. It was firstly equipped with a reconnaissance camera, while in 2002 it was renamed as "MQ-1"; "R" for "Reconnaissance" and "M" for "Multi-role" (Giese et al., 2013).

Civilian drones

Considering the technological progress of military drones over the past years, government agencies started testing these technologies also for other purposes including border surveillance, castaway rescue, and firefighting response. In parallel, the landscape was shifted toward commercial exploitation. During approximately the last 20 years, taking advantage of the fast-paced innovation (and often inspired by insects), drones tend to become smaller and smaller. Moreover, the flight duration and the quality of the embedded cameras and sensors are continuously improved at rapid speeds. As a final note, drones have become more accessible to the masses and are being used for applications that could not be imagined before some years. For instance, the initial surveillance applications for military purposes have also been extended to other domains, like livestock monitoring and asset protection. Moreover, drones are broadly utilized for environmental monitoring, response to humanitarian disasters, surveying and mapping, as well as engineering and construction. Delivery applications have also been emerged, such as healthcare, aiming at delivery of medicines and vaccines to remote regions. Also, corporations like DHL (Dalsey, Hillblom and Lynn) and Amazon are using commercial drones for delivery activities. Finally, drones constitute an integral part of precision agriculture and have driven the revolution of this domain in recent years. In subsection "Classification based on application," some civilian applications of drones will be described along with representative examples.

Classification of drones

Classification based on weight and size

A first categorization of drones can be done on the basis of their weight. According to Arjomandi et al. (2006), drones can be classified into "micro" (w ≤ 5 kg), "light" (5 kg ≤ w ≤ 50 kg), "medium" (50 kg ≤ w ≤ 200 kg), "heavy" (200 kg ≤ w ≤ 2000 kg), and "superheavy" (w > 2000 kg). Similar classifications have been proposed in Brooke-Holland (2012), Watts et al. (2012), Weibel and Hansman (2004), and Homainejad and Rizos (2015). In Hassanalian and Abdelkefi (2017a), a classification was suggested in reference with both weight and wingspan of drones ranged from vast fixed-wing UAVs to "smart dust" consisting of several micro-electromechanical system (MEMS) (Fig. 1.2).

Classification based on the aerodynamic features

Roughly speaking, drones may either have fixed or flapping wings or utilize rotors or copters to fly. Also, hybrid drones that combine the

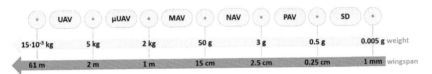

Figure 1.2 Classification of drones based on weight and size. *UAV*, Unmanned air vehicle; *μUAV*, micro unmanned air vehicle; *MAV*, micro air vehicle; *NAV*, nano air vehicle; *PAV*, pico air vehicle; *SD*, smart dust.

operating principles of the above aerial vehicles are increasingly gaining attention.

Fixed wing

As the term implies, this type of aerial vehicles utilizes fixed wings similar to airplanes. Apart from the rigid wings, also fuselage and tails using a propeller and motor make up these drones. Because of the relatively low required power, since they do not consume energy to maintain height, they can cover long distances at high speeds. Their greatest weakness, however, is that they only move forward and, hence, they are not able to hover in one spot or to land at a specific point. In particular, a runway is usually implemented for landing them or even a net for the purpose of safely catching them without being damaged. Besides, when using fixed-wing drones there is always the danger of hitting obstacles (Dharmawan et al., 2019), since they cannot stop or reduce speed below a flying limit, as mentioned above. In a nutshell, fixed-wing drones are ideal for large-scale aerial mapping and long-range inspections in a broad spectrum of operational environments, such as desert, maritime, arctic, jungle, and agricultural fields (Hassanalian and Abdelkefi, 2017b; Hakim et al., 2021).

Flapping wing

Inspired from nature's original flying mechanics, flapping-wing drones have been fabricated (Billingsley et al., 2021; Hassanalian et al., 2017; Chen and Zhang, 2019). This type of drones is usually applicable in micro air vehicles (MAVs), nano air vehicles (NAVs), and pico air vehicles (PAVs). More specifically, the design of flapping-wing drones is being inspired from birds or bats (MAVs), insects (PAVs), or from creatures between big insects and small birds (NAVs), like dragonflies and hummingbirds. Bio-inspired drones take advantage of the progress in micro-fabrication technology and dynamics of birds' and insects' flight (Wood, 2007). A representative example is the "Bat Bot" drone that mimics the flexible way that bats fly by implementing ultra-thin membrane wings made from silicon (Ramezani et al., 2017). The hovering ability of insects in conjunction with their unique maneuverability advantages, including rapid turning and abrupt acceleration, is an ideal paradigm for drone research (Orlowski and Girard, 2012). Usually, three configurations exist, namely monoplane (a single pair of flapping wings), biplane (two superimposed pairs of flapping wings), and combination of them (two sets of flapping wings operating independently, with one behind the other). One of the disadvantages of drones utilizing flapping wings is the high-power required for the flapping operation.

Rotary wing

There are various types of drones with rotary wings having from 1 to 12 motors, named on the basis of the number of rotors they utilize. In the relative literature, they are usually referred to (with or without the dash) as:

1. Mono-copter or single-copter (1 rotor);
2. Twin-copter (2 rotors);
3. Tri-copter (3 rotors);
4. Quad-copter (4 rotors);
5. Penta-copter (5 rotors);
6. Hexa-copter (6 rotors);
7. Octo-copter (8 rotors);
8. Deca-copter (10 rotors);
9. Dodeca-copter (12 rotors).

The most commonly used drones with rotary wings are the quad-copters and hexa-copters (Hassanalian and Abdelkefi, 2017a; Cai et al., 2014). In general, multicopters are, most of the time, the cheapest option in the market. They have high maneuverability and the capability to hover as well as fly in every direction, making them exemplary aerial vehicles for inspecting hard-to-reach regions. As far as mono-copters are concerned, they use a tail to control their flight like conventional helicopters. They are usually preferred when long distances must be surveyed or carrying heavier weights. Nevertheless, they are not as easy to pilot as multicopters, while they are considered not to be safe enough, because of their larger spinning blades. Finally, although multicopters do have their shortcomings, they are constantly getting better and better resulting in improving their overall flight times and reducing their energy consumption.

Cyclocopter

Cyclocopters utilize a cyclo-rotor as a rotor wing as a means of providing lift, control, and propulsion. The wing is similar to a paddle wheel having airfoil blades instead of paddles (Yun et al., 2004). Cyclocopters, like helicopters, can vertically takeoff, hover, and land, a feature always being an engineering challenge for the sake of safety. Their engineering design makes them less influenced by wind gusts with better maneuverability and altitude limit enabling them for surveying inaccessible regions after natural disasters or even other planets for exploration (Husseyin and Warmbrodt, 2016). An additional advantage of cyclocopters is that they tend to better perform when scaled down. More specifically, the vortex created by the cyclo-rotors gets proportionally more powerful with decreasing the size, thus, making cyclocopters one of the leading candidates for miniaturized drones.

Hybrid

Toward the direction of getting a best-of-both-worlds characteristics set, hybridization has been proposed. For example, fixed-wing drones have been designed and fabricated that also deploy rotors. As a consequence, they can combine the benefits of long flight times of fixed-wing counterparts with the additional capability of hovering in one place or performing vertical takeoff and land (Gunarathna and Munasinghe, 2018). Furthermore, there exist drones using both flapping wings and fixed wings. Through this hybrid model, a more aerodynamically balanced platform can be developed with a better efficiency (Jones et al., 2005). The term "hybrid," similarly to terminology used in car industry, is also used to describe drones that combine energy sources for powering drones, such as lithium batteries with wind—solar power generation systems (Du et al., 2019) or solar cells (Bronz et al., 2009; Hassanalian et al., 2014). Concerning the case of micro drones, energy can be even produced from the wasted mechanical energy of flapping motion and morphing for the sake of operating many on-board sensors (Abdelkefi and Ghommem, 2013).

Classification based on application

As elaborated in the recent review studies of Labib et al. (2021) and Maghazei and Netland (2020), drones can be simply classified into three categories depending on their function for which they are intended, namely: (1) perceive, (2) act, and (3) perceive and act.

Perceive

In the "perceive" category, the main goal of drones is to perceive the physical world's features by collecting useful data. Examples belonging in this category are search operations, traffic, asset and environmental monitoring, security surveillance, infrastructure inspection, and remote sensing for precision agriculture applications.

Search operations during large-scale disasters

In the case of a large-scale disaster, including tsunamis, earthquakes, hurricanes, floods, and fires, the most vital issue is how to preserve human lives. As a matter of fact, the search operations must be conducted in a both effective and timely manner. Nowadays, increasing effort has been observed in exploiting a network of drones for efficient reconnaissance, mapping, and evacuation support. This means better awareness of the number of people needing help, the extent of the damage to the natural environment or existing structures, and the conditions of the transport infrastructure (Erdelj et al., 2017). Additionally, drones can be used for supporting the communication infrastructure, for instance, through connecting mobile phones to the closest radio access networks or as stand-alone communication systems (Câmara, 2014). The major challenges for the above applications are the low energy efficiency of drones and the adverse conditions they may face with during their flight like unpredictable air drafts.

Environmental monitoring

A useful set of applications of drones that is correlated with the above emergency operations is forecasting and early warning through environmental monitoring. Moreover, taking into account that societies are becoming more and more aware of air pollution and the resultant health problems, drones can constitute valuable assets toward monitoring it (Thomazella et al., 2018). Common pollutants that have to be measured are usually carbon monoxide (CO), nitrogen dioxide (NO_2), sulfur dioxide (SO_2), lead (Pb), and particulate matter (usually $PM_{2.5}$ and PM_{10}). Due to their fixed position, conventional monitoring systems demonstrate low spatial coverage. In contrast, by taking advantage of drones' flexibility, the target places can be easily reached on demand (Dugdale et al., 2018). Depending on the on-board sensors, drones can also monitor other parameters such as noise, temperature, wind parameters, and humidity.

Traffic monitoring

A consequence of the increasing traffic congestion problems in cities is the need for more efficient information toward real-time management of traffic. This calls for accurate information regarding not only roadway conditions but emergency cases as well, which can result in traffic congestion, such as car accidents and bad road conditions. Among the proposed approaches to address the problems related to traffic monitoring, like radar, video, and ultra-sound technologies, drones appear to be a viable alternative because of their nonrestricted mobility and wide view range and low cost (Kanistras et al., 2013). In contrast with the traditional stationary systems, drones provide higher flexibility, as they are not located at fixed positions and, thus, can be used where required to get traffic data or capture individual vehicles' movements. However, this is challenging enough, since real-time road monitoring requires both suitable on-board vision sensors to collect the information and data processing along with challenges related to motion of background (Najiya and Archana, 2018).

Infrastructure inspection

Drones are increasingly deployed for asset inspection, including photovoltaic panels (Niccolai et al., 2019) and wind farms (Shafiee et al., 2021). The maintenance of these plants plays a key role in energy production. Consequently, drones have the potential to decrease the cost through reducing the number of times needed for the personnel to travel to the solar panels or wind turbines and climb up them with the heavy equipment. In addition, drones can shorten the downtime duration required to detect defects and gather diagnostic information. A remarkable application of drones is also railway track inspections so as to monitor the conditions toward ensuring safety against potential accidents (Mammeri et al., 2021). Interestingly, the rapid progress of machine vision and deep learning techniques has boosted the use of drones for the purpose of automatically detecting railway track defects (Bojarczak and Lesiak, 2021).

Security surveillance

By exploiting the progress in machine vision technologies, security surveillance based on drones can provide large amounts of videos and images, which can be utilized to identify the region or the person of interest (Nguyen et al., 2021; Gassara and Bouassida Rodriguez, 2021). Furthermore, safety threats are present within industrial environments requiring real-time and trustworthy situation awareness. Toward that direction, routine drones' inspections have the potential of preventing endangerment of employers and providing emergency response.

Remote sensing for precision agriculture applications

Precision agriculture takes into consideration both temporal and spatial variability with the intention of improving agricultural production by simultaneously bearing in mind sustainability (Khun et al., 2021; Lampridi et al., 2019). As a means of achieving these goals, drones have been adopted for the purpose of collecting useful information from several sources, which can result in useful conclusions pertaining to acting at the right place and time with the right quantity (Khun et al., 2021; Mulla, 2013; Tagarakis et al., 2022). Among the most common technologies in precision agriculture, remote sensing can be very useful for a lot of applications. Remote sensing includes an assortment of sensors, mainly optical. For instance, images taken from sensors at various wavelengths can be used to provide useful diagnostic information in the direction of optimal crop management, yield prediction, and protection of the environment (Angelopoulou et al., 2019; Moshou et al., 2011; Pantazi et al., 2019). Furthermore, thermal data are suitable for examining the water status, whereas spectral data are used to detect plant diseases (Radoglou-Grammatikis et al., 2020). Aggregation of thermal and multispectral data is also utilized for evaluating the health status and water stress (Katsigiannis et al., 2016). As the implementation of drones in farming has become very common in the last years, several review studies exist in the relative literature such as Tagarakis et al. (2022), Esposito et al. (2021), Panday et al. (2020), and Anagnostis et al. (2021a). As a general conclusion, the most common applications involve (1) weed mapping, (2) monitoring of the vegetation health and growth, and (3) irrigation management.

Act
This category focuses on acting in the physical world. Indicative examples are spraying drones for crops' fertilization, as a means of providing them with the required nutrients, targeted application of pesticides, and disinfection of public surfaces against the recent COVID-19 coronavirus. Applications are also extended to the autonomous on-demand delivery of medical supplies, small packages, and food as well as emergency services.

Pest control and fertilization for precision agriculture applications
As has already been mentioned, a sector that has benefited from several types of drones is agriculture. In particular, drones have been adopted in precision agriculture owing to their ability to carry out informed low-altitude operations (Moses-Gonzales and Brewer, 2021) for gathering data at high resolution and in a cost-effective manner for detecting pests (e.g., weeds and arthropods) (Albetis et al., 2017; Huang et al., 2018). A remarkable evolution of drones pertaining to this application is an integrated system for remote sensing and mapping of the pest-infested zone

and precision spraying. Ideally, although a single drone conducting these tasks is desirable, both seeking pests and flying by carrying a heavy tank with a pesticide solution considerably reduces the battery charge duration (Hunter et al., 2020). Usually, small drones are deployed for pest detection and other drones for aerial spraying directly flying to the infected areas. Apart from pesticides, also fertilizers in a liquid form can be sprayed from spraying drones in a targeted way resulting in less inputs for the benefit of both cost saving as well as environmental and human health protection (Hentschke et al., 2018; Benos et al., 2020).

Spraying drones are a characteristic type of drones belonging in "act" category and constitute a kneading of hardware, software, mechanics, and fluid dynamics (Rahman et al., 2021). The main components of spraying drones are usually pressure nozzles, a spraying controller, a tank with chemical solution, a miniature diaphragm pump, a hall-flow sensor, and a field-map reading system (Velusamy et al., 2022). The fluid existing within the tank flows to the nozzles, where it splits into small droplets under pressure, while a motor with a pump is utilized to produce the required pressure. Additionally, the hall-flow sensor is used by the spraying controller for appraising the fluid flow inside the system (Lou et al., 2018). Finally, spraying drones can vary on the basis of payload, speed, and number of nozzles.

Fighting the COVID-19 pandemic

An unexpected factor that has impacted almost all aspects of our everyday life and tested the resilience of several sectors was the COVID-19 pandemic outbreak (Bochtis et al., 2020). The possibility of using drones to contain the spread of the coronavirus was investigated and implemented (Restás, 2022). An indicative application is using spraying drones as an alternative technique for the disinfection of public surfaces, with the disinfectant liquid flow rate along with the altitude and flight speed during the mission being key parameters (Restás et al., 2021; González Jorge et al., 2021). Other applications of drones related to COVID-19 are the aerial surveillance to screen out febrile patient in open public areas by using drones with thermal cameras or face mask detection in public areas (Almalki et al., 2021). Furthermore, some countries, like China, Portugal, Spain, Italy, Rwanda, and the United States, are using drones with QR code flags or loudspeakers to monitor citizens' travel history (through a mobile app) (Lu) or to provide announcements for staying at home, keeping social distances, and wearing a face mask (Huaxia Portugal).

On-demand delivery

Using drones for the purpose of light payload delivery offers an advantageous solution to various transportation problems, since drones do not contribute to traffic congestion and are both clean and fast (Liu, 2019). Applications are extended to meal delivery, or rapid parcel delivery of small packages to customers from corporations like Amazon in the United States and DHL post service in Germany. Remarkably, drones have also been used for delivering medical supplies, such as medicines, vaccines and blood, to disaster areas or regions having poor transport networks or extreme traffic congestion (Hii et al., 2019). In Rwanda, for instance, drones are deployed for delivering blood by throwing the package with a parachute on the ground (Ackerman and Koziol, 2019).

As noted in Stolaroff et al. (2018), drone deliveries are more environment-friendly than the traditional methods and have the potential to complement the latter. Overall, in addition to the obvious challenges pertaining to this technology, including the battery life, payload, and operational range, drones are vulnerable to unpredictable weather conditions. Besides, many hardware and software costs are still high enough, leading to considerably high investment costs. Drone delivery should also solve issues associated with finding a safe landing area, possible hacking, regulation, and social acceptance.

Perceive and act

Arguably, the most challenging application of drones is to both perceive and act, as a higher degree of autonomy is required for this purpose. This demanding category contains less applications, although the advancement in the technologies of sensors and drones has the potential to open up new horizons for drones' applications. Overall, the applications described above tend to be automated with drones carrying out more tasks involving both perceiving and acting. Indicatively, an integral and challenging part of the search operations described above is also rescue. In addition, researchers and manufacturers are trying to fabricate localized solutions for twofold real-time pest management by combining finding out of pest hotspots and fighting them with a single drone.

Search and rescue operations during large-scale disasters

Despite the limitations related to payload weight, drones have the potential to detect people in need and keep them alive by delivering essential supplies, especially in the case of a damaged transport infrastructure or hard-to-reach regions. An indicative example is the drone called "Pars" (Smith). This is a remotely controlled drone that can identify castaways struggling in the sea and drop up to three life rings. Furthermore, combined approaches have been proposed. For example, in Erdelj et al. (2017) it is suggested a system of a fixed-wing drone for quickly surveying the affected region and five quad-copters to approach the critical spots. According to Erdelj et al. (2017), disaster management through drones can be comprised by the following three phases:

1. Predisaster preparedness: Survey associated events preceding the natural disaster.
2. Disaster assessment: Real-time situation awareness via drones logistical planning.
3. Disaster response and recovery: Support of search and rescue operations and building communications links.

In general, as these phases progress, the drones' effectiveness increases. In contrast, the static wireless sensor network (usually abbreviated as WSN) is no longer sustainable.

Identification and pest destroying for precision agriculture applications

Nowadays, efforts are put as a means of developing innovative drones combining the capabilities of both identifying and destroying pests in real time, taking also advantage of the advancements in machine learning solutions dedicated to pests identification (Liakos et al., 2018; Benos et al., 2021; Anagnostis et al., 2020, 2021b). However, as payload and flight time are limiting factors, it is very challenging to create an aerial vehicle toward this direction. Nevertheless, as highlighted in Iost Filho et al. (2020), if large areas need to be covered, a swarm of communicating drones seems to be most efficient.

Discussion and main conclusions

In conclusion, considering that drones vary in size, weight, power, load, endurance time, and purpose, there exist several taxonomic approaches. For instance, drones can serve for either military or civilian purposes. The former was the first field of application and constituted the forerunners and the motivation for shifting the landscape toward civilian exploitation. Currently, drones are more and more accessible to the masses and are used either for professional purposes or aerial photography and videography. Furthermore, taking into account the metrics of size and weight of drones, various classifications have been proposed. In this chapter, the categorization of drones according to Arjomandi et al. (2006) and Hassanalian and Abdelkefi (2017a) was briefly described.

Based on the aerodynamic features, drones are commonly classified into fixed-wing, flapping-wing, rotary-wing, and hybrid drones. In short, fixed-wing drones are regarded as the easiest ones in terms of design and fabrication and are utilized for covering large regions at high speed. However, one of their major drawbacks is their inability to hover, while flying in indoor areas is very difficult. Inspired by insects and birds, flapping-wing drones can hover and fly in every direction, in contrast with the fixed-wing ones that can only move forward. These advantages are also present in the case of drones with rotary wings, rendering them ideal for approaching hard-to-reach regions. Nevertheless, both flapping-wing and rotary-wing drones require high energy consumption. A special category of rotary-wing drones is cyclocopters, which are less influenced by wind gusts and are considered suitable for miniaturized drones. Finally, hybrid drones have also been fabricated combining the best features of each category as well as energy sources like batteries with solar cells. Also the taxidermy bodies of animals have been used as structural components of drones along with other elements including sensors and electrical batteries as well as live birds or insects controlled through embodied electrical chips (Hassanalian and Abdelkefi, 2017a).

As far as the applications of drones are concerned, drones can provide tactical surveillance, save human lives, contribute to maintenance of infrastructure, build communication links at hard-to-reach places, and modernize agriculture by conserving resources, improving yield, and avoiding waste. A common denominator that can be used for the purpose of grouping the existing applications of drones is the tasks for which they are programmed to carry out. To that end, drones can either perceive or act. Considering the advancement in various scientific fields, including artificial intelligence, machine vision, sensors, and information technology, the drones can also combine both the capabilities of perceiving and acting. However, more research is needed toward that direction, owing to the various challenges that must be overcome.

Nowadays, new materials are investigated for constructing components of drones, which are characterized by lighter weight, while 3D-printing is increasingly deployed allowing for faster fabrication of drones at decreased cost and consequently commercially available in lower prices.

Additionally, new propulsion technologies are examined. The most common way is the use of electric motors because of their low vibration and low energy consumption. For the purpose of enhancing the flight endurance or supporting the operation of some sensors, also renewable energy sources have been used, such as solar cells.

Using a swarm of drones is considered to be a "game-changer" in many areas, including military, precision agriculture, and search and rescue operations, that has the potential of reducing the time and increasing the derived data for any application. As a matter of fact, it is regarded as the next key leap forward for drones taking advantage of the progress in IoT. A drone swarm does not have to be made of the same kind of drones; in principle, it can deploy any heterogenous type of drone based on the application at hand. This can provide more flexibility. Military applications seem to be the first to exploit the benefits of swarms, whereas civilian applications are not expected to be far behind them.

As a last remark, even though the promised benefits by the adoption of drones cannot be disputed, some key challenges should be addressed, apart from technological aspects, ranging from privacy, data protection, security, safety, regulation, to social acceptance (Benos et al., 2022).

References

Abdelkefi, A., Ghommem, M., 2013. Piezoelectric energy harvesting from morphing wing motions for micro air vehicles. Theor. Appl. Mech. Lett. 3, 52004. Available from: https://doi.org/10.1063/2.1305204.

Ackerman, E., Koziol, M., 2019. The blood is here: zipline's medical delivery drones are changing the game in Rwanda. IEEE Spectr. 56, 24–31. Available from: https://doi.org/10.1109/MSPEC.2019.8701196.

Albetis, J., Duthoit, S., Guttler, F., Jacquin, A., Goulard, M., Poilvé, H., et al., 2017. Detection of flavescence dorée grapevine disease using unmanned aerial vehicle (UAV) multispectral imagery. Remote Sens. 9.

Almalki, F.A., Alotaibi, A.A., Angelides, M.C., 2021. Coupling multifunction drones with AI in the fight against the coronavirus pandemic. Computing . Available from: https://doi.org/10.1007/s00607-021-01022-9.

Anagnostis, A., Asiminari, G., Papageorgiou, E., Bochtis, D., 2020. A convolutional neural networks based method for anthracnose infected walnut tree leaves identification. Appl. Sci. 10, 469. Available from: https://doi.org/10.3390/app10020469.

Anagnostis, A., Tagarakis, A.C., Kateris, D., Moysiadis, V., Sørensen, C.G., Pearson, S., et al., 2021a. Orchard mapping with deep learning semantic segmentation. Sensors 21. Available from: https://doi.org/10.3390/S21113813.

Anagnostis, A., Tagarakis, A.C., Asiminari, G., Papageorgiou, E., Kateris, D., Moshou, D., et al., 2021b. A deep learning approach for anthracnose infected trees classification in walnut orchards. Comput. Electron. Agric. 182, 105998. Available from: https://doi.org/10.1016/j.compag.2021.105998.

Angelopoulou, T., Tziolas, N., Balafoutis, A., Zalidis, G., Bochtis, D., 2019. Remote sensing techniques for soil organic carbon estimation: a review. Remote Sens. 11. Available from: https://doi.org/10.3390/rs11060676.

Anush Lakshman, S., Ebenezer, D., 2021. Integration of internet of things and drones and its future applications. Mater. Today Proc. 47, 944—949. Available from: https://doi.org/10.1016/j.matpr.2021.05.039.

Beg, A., Qureshi, A.R., Sheltami, T., Yasar, A., 2021. UAV-enabled intelligent traffic policing and emergency response handling system for the smart city. Pers. Ubiquitous Comput. 25, 33—50. Available from: https://doi.org/10.1007/s00779-019-01297-y.

Benos, L., Bechar, A., Bochtis, D., 2020. Safety and ergonomics in human-robot interactive agricultural operations. Biosyst. Eng. 200, 55—72. Available from: https://doi.org/10.1016/j.biosystemseng.2020.09.009.

Benos, L., Sørensen, C.G., Bochtis, D., 2022. Field deployment of robotic systems for agriculture in light of key safety, labor, ethics and legislation issues. Curr. Robot. Rep. Available from: https://doi.org/10.1007/s43154-022-00074-9.

Benos, L., Tagarakis, A.C., Dolias, G., Berruto, R., Kateris, D., Bochtis, D., 2021. Machine learning in agriculture: a comprehensive updated review. Sensors 21. Available from: https://doi.org/10.3390/S21113758.

Billingsley, E., Ghommem, M., Vasconcellos, R., Abdelkefi, A., 2021. On the aerodynamic analysis and conceptual design of bioinspired multi-flapping-wing drones. Drones 5. Available from: https://doi.org/10.3390/drones5030064.

Bochtis, D., Benos, L., Lampridi, M., Marinoudi, V., Pearson, S., Sørensen, C.G., 2020. Agricultural workforce crisis in light of the COVID-19 pandemic. Sustainability 12. Available from: https://doi.org/10.3390/su12198212.

Bojarczak, P., Lesiak, P., 2021. UAVs in rail damage image diagnostics supported by deep-learning networks. Open. Eng. 11, 339—348. Available from: https://doi.org/10.1515/eng-2021-0033.

Bronz, M., Moschetta, J.M., Brisset, P., Gorraz, M., 2009. Towards a Long Endurance MAV. Int. J. Micro Air Veh. 1, 241—254. Available from: https://doi.org/10.1260/175682909790291483.

Brooke-Holland, L., 2012. Unmanned Aerial Vehicles (Drones): An Introduction. House of Commons Library, UK.

Cai, G., Dias, J., Seneviratne, L., 2014. A survey of small-scale unmanned aerial vehicles: recent advances and future development trends. Unmanned Syst. 2, 175—199. Available from: https://doi.org/10.1142/S2301385014300017.

Chechin, D.G., Artamonov, A.Y., Bodunkov, N.Y., Kalyagin, M.Y., Shevchenko, A.M., Zhivoglotov, D.N., 2021. Development of an unmanned aerial vehicle to study atmospheric boundary-layer turbulent structure. J. Phys. Conf. Ser. 1925, 12068. Available from: https://doi.org/10.1088/1742-6596/1925/1/012068.

Chen, C., Zhang, T., 2019. A review of design and fabrication of the bionic flapping wing micro air vehicles. Micromachines 10. Available from: https://doi.org/10.3390/mi10020144.

Cromwell, C., Giampaolo, J., Hupy, J., Miller, Z., Chandrasekaran, A., 2021. A systematic review of best practices for UAS data collection in forestry-related applications. Forests 12. Available from: https://doi.org/10.3390/f12070957.

Arjomandi, A., Agostino, S., Mammone, M., Nelson, M., Zhou, T., 2006. Classification of Unmanned Aerial Vehicle, Report for Mechanical Engineering class. University of Adelaide, Adelaide, Australia.

Câmara, D., 2014. Cavalry to the rescue: drones fleet to help rescuers operations over disasters scenarios. In: Proc. of the 2014 IEEE Conference on Antenna Measurements & Applications (CAMA), pp. 1—4.

Dharmawan, A., Putra, A.E., Tresnayana, I.M., Wicaksono, W.A., 2019. The obstacle avoidance system in a fixed-wing UAV when flying low using LQR method. In: Proc. of the 2019 International Conference on Computer Engineering, Network, and Intelligent Multimedia (CENIM), pp. 1–7.

Dugdale, S., Kelleher, C., Malcolm, I., Hannah, D.M., 2018. Utility of drone-based thermal imaging for mapping river temperature heterogeneity. In: Proc. of the AGU Fall Meeting Abstracts, Vol. 2018, NS43C-0852.

Du, Y., Lei, W., Fei, W., 2019. Design and implementation of a wind solar hybrid power generation system. In: Proc. of the 4th International Conference on Advances in Energy and Environment Research (ICAEER), Shanghai.

Erdelj, M., Natalizio, E., Chowdhury, K.R., Akyildiz, I.F., 2017. Help from the sky: leveraging UAVs for disaster management. IEEE Pervasive Comput. 16, 24–32. Available from: https://doi.org/10.1109/MPRV.2017.11.

Esposito, M., Crimaldi, M., Cirillo, V., Sarghini, F., Maggio, A., 2021. Drone and sensor technology for sustainable weed management: a review. Chem. Biol. Technol. Agric. 8, 18. Available from: https://doi.org/10.1186/s40538-021-00217-8.

Fotouhi, A., Qiang, H., Ding, M., Hassan, M., Giordano, L.G., Garcia-Rodriguez, A., et al., 2019. Survey on UAV cellular communications: practical aspects, standardization advancements, regulation, and security challenges. IEEE Commun. Surv. Tutor. (21), 3417–3442. Available from: https://doi.org/10.1109/COMST.2019.2906228.

Ganesh, S., Gopalasamy, V., Shibu, N.B.S., 2021. Architecture for drone assisted emergency ad-hoc network for disaster rescue operations. In: Proc. of the 2021 International Conference on COMmunication Systems & NETworkS (COMSNETS), pp. 44–49.

Gassara, A., Bouassida Rodriguez, I., 2021. Describing correct UAVs cooperation architectures applied on an anti-terrorism scenario. J. Inf. Secur. Appl. 58, 102775. Available from: https://doi.org/10.1016/j.jisa.2021.102775.

Giese, S., Carr, D., Chahl, J. 2013. Implications for unmanned systems research of military UAV mishap statistics. In: Proc. of the 2013 IEEE Intelligent Vehicles Symposium (IV), pp. 1191-1196.

Goh, K.C.W., Ng, R.B.C., Wong, Y.-K., Ho, N.J.H., Chua, M.C.H., 2021. Aerial filming with synchronized drones using reinforcement learning. Multimed. Tools Appl. 80, 18125–18150. Available from: https://doi.org/10.1007/s11042-020-10388-5.

González Jorge, H., de Santos, L.M., Fariñas Álvarez, N., Martínez Sánchez, J., Navarro Medina, F., 2021. Operational study of drone spraying application for the disinfection of surfaces against the COVID-19 pandemic. Drones 5. Available from: https://doi.org/10.3390/drones5010010.

Gschwindt, M., Camci, E., Bonatti, R., Wang, W., Kayacan, E., Scherer, S., 2019 Can a robot become a movie director? Learning artistic principles for aerial cinematography. In: Proc. of the 2019 IEEE/RSJ International Conference on Intelligent Robots and Systems (IROS), pp. 1107–1114.

Guimarães, N., Pádua, L., Marques, P., Silva, N., Peres, E., Sousa, J.J., 2020. Forestry remote sensing from unmanned aerial vehicles: a review focusing on the data, processing and potentialities. Remote Sens. 12. Available from: https://doi.org/10.3390/rs12061046.

Gunarathna, J.K., Munasinghe, R., 2018. Development of a quad-rotor fixed-wing hybrid unmanned aerial vehicle. In: Proc. of the 2018 Moratuwa Engineering Research Conference (MERCon), pp. 72–77.

Hakim, M.L., Pratiwi, H., Nugraha, A.C., Yatmono, S., Wardhana, A.S.J., Damarwan, E. S., et al., 2021. Development of unmanned aerial vehicle ( ) fixed-wing for monitoring, mapping and dropping applications on agricultural land. J. Phys. Conf. Ser. 2111, 12051. Available from: https://doi.org/10.1088/1742-6596/2111/1/012051.

Hassanalian, M., Radmanesh, M., Sedaghat, A., 2014. Increasing flight endurance of MAVs using multiple quantum well solar cells. Int. J. Aeronaut. Space Sci. 15, 212−217. Available from: https://doi.org/10.5139/ijass.2014.15.2.212.

Hassanalian, M., Abdelkefi, A., Wei, M., Ziaei-Rad, S., 2017. A novel methodology for wing sizing of bio-inspired flapping wing micro air vehicles: theory and prototype. Acta Mech. 228, 1097−1113. Available from: https://doi.org/10.1007/s00707-016-1757-4.

Hassanalian, M., Abdelkefi, A., 2017a. Classifications, applications, and design challenges of drones: a review. Prog. Aerosp. Sci. 91, 99−131. Available from: https://doi.org/ 10.1016/j.paerosci.2017.04.003.

Hassanalian, M., Abdelkefi, A., 2017b. Design, manufacturing, and flight testing of a fixed wing micro air vehicle with Zimmerman planform. Meccanica 52, 1265−1282. Available from: https://doi.org/10.1007/s11012-016-0475-2.

Hentschke, M., de Freitas, E., Hennig, C.H., da Veiga, I.C., 2018. Evaluation of altitude sensors for a crop spraying drone. Drones 2. Available from: https://doi.org/10.3390/ drones2030025.

Hii, M.S.Y., Courtney, P., Royall, P.G., 2019. An evaluation of the delivery of medicines using drones. Drones 3. Available from: https://doi.org/10.3390/drones3030052.

Homainejad, N., Rizos, C., 2015. Application of multiple categories of unmanned aircraft systems (UAS) in different airspaces for bushfire monitoring and response. Int. Arch. Photogramm. Remote Sens. Spat. Inf. Sci. 55−60. Available from: https://doi.org/ 10.5194/isprsarchives-XL-1-W4-55-2015.

Huang, Y.; Reddy, K.N.; Fletcher, R.S.; Pennington, D. UAV low-altitude remote sensing for precision weed management. Weed Technol. 2018, 32, 2−6, doi: 10.1017/ wet.2017.89.

Huang, H., Savkin, A.V., Huang, C., 2021. Decentralized autonomous navigation of a UAV network for road traffic monitoring. IEEE Trans. Aerosp. Electron. Syst. 57, 2558−2564. Available from: https://doi.org/10.1109/TAES.2021.3053115.

Huaxia Portugal Starts Using Talking Drones to Tell People to Stay at Home.

Hunter 3rd, J.E., Gannon, T.W., Richardson, R.J., Yelverton, F.H., Leon, R.G., 2020. Integration of remote-weed mapping and an autonomous spraying unmanned aerial vehicle for site-specific weed management. Pest. Manag. Sci. 76, 1386−1392. Available from: https://doi.org/10.1002/ps.5651.

Husseyin, S., Warmbrodt, W.G., 2016. Design Considerations for a Stopped-Rotor Cyclocopter for Venus Exploration. NASA International Space University Intern, Aeromechanics Office, NASA Ames Research Center: Moffet, Field, CA, USA.

Iost Filho, F.H., Heldens, W.B., Kong, Z., de Lange, E.S., 2020. Drones: innovative technology for use in precision pest management. J. Econ. Entomol. 113, 1−25. Available from: https://doi.org/10.1093/jee/toz268.

Jones, K.D., Bradshaw, C.J., Papadopoulos, J., Platzer, M.F., 2005. Bio-inspired design of flapping-wing micro air vehicles. Aeronaut. J. 109, 385−393. Available from: https:// doi.org/10.1017/S0001924000000804.

Kaivosoja, J., 2022. Future possibilities and challenges for UAV-based imaging development in smart farming. In: Lipping, T., Linna, P., Narra, N. (Eds.), New Developments and Environmental Applications of Drones. Springer International Publishing, Cham, pp. 109−119.

Kanistras, K., Martins, G., Rutherford, M.J., Valavanis, K.P., 2013 A survey of unmanned aerial vehicles (UAVs) for traffic monitoring. In: Proc. of the 2013 International Conference on Unmanned Aircraft Systems (ICUAS), pp. 221−234.

Katsigiannis, P., Misopolinos, L., Liakopoulos, V., Alexandridis, T.K., Zalidis, G., 2016 An autonomous multi-sensor UAV system for reduced-input precision agriculture applications. In: Proc. of the 2016 24th Mediterranean Conference on Control and Automation (MED), pp. 60−64.

Khan, S.I., Qadir, Z., Munawar, H.S., Nayak, S.R., Budati, A.K., Verma, K.D., et al., 2021. UAVs path planning architecture for effective medical emergency response in future networks. Phys. Commun. 47, 101337. Available from: https://doi.org/10.1016/j.phycom.2021.101337.

Khun, K., Tremblay, N., Panneton, B., Vigneault, P., Lord, E., Cavayas, F., et al., 2021. Use of oblique RGB imagery and apparent surface area of plants for early estimation of above-ground corn biomass. Remote Sens. 13.

Konert, A., Balcerzak, T., 2021. Military autonomous drones (UAVs) − from fantasy to reality. Legal and ethical implications. Transp. Res. Procedia 59, 292−299. Available from: https://doi.org/10.1016/j.trpro.2021.11.121.

Labib, N.S., Brust, M.R., Danoy, G., Bouvry, P., 2021. The rise of drones in internet of things: a survey on the evolution, prospects and challenges of unmanned aerial vehicles. IEEE Access. 9, 115466−115487. Available from: https://doi.org/10.1109/ACCESS.2021.3104963.

Lambey, V., Prasad, A.D., 2021. A review on air quality measurement using an unmanned aerial vehicle. Water Air Soil Pollut. 232, 109. Available from: https://doi.org/10.1007/s11270-020-04973-5.

Lampridi, M., Sørensen, C., Bochtis, D., 2019. Agricultural sustainability: a review of concepts and methods. Sustainability 11, 5120. Available from: https://doi.org/10.3390/su11185120.

Liakos, K., Busato, P., Moshou, D., Pearson, S., Bochtis, D., Liakos, K.G., et al., 2018. Machine learning in agriculture: a review. Sensors 18, 2674. Available from: https://doi.org/10.3390/s18082674.

Liu, Y., 2019. An optimization-driven dynamic vehicle routing algorithm for on-demand meal delivery using drones. Comput. Oper. Res. 111, 1−20. Available from: https://doi.org/10.1016/j.cor.2019.05.024.

Li, X., Li, Z., Wang, H., Li, W., 2021. Unmanned aerial vehicle for transmission line inspection: status, standardization, and perspectives. Front. Energy Res. 9, 336. Available from: https://doi.org/10.3389/fenrg.2021.713634.

Lou, Z., Xin, F., Han, X., Lan, Y., Duan, T., Fu, W., 2018. Effect of unmanned aerial vehicle flight height on droplet distribution, drift and control of cotton aphids and spider mites. Agronomy 8. Available from: https://doi.org/10.3390/agronomy8090187.

Lo, L.-Y., Yiu, C.H., Tang, Y., Yang, A.-S., Li, B., Wen, C.-Y., 2021. Dynamic object tracking on autonomous UAV system for surveillance applications. Sensors 21. Available from: https://doi.org/10.3390/s21237888.

Lu, D., China Is Using Mass Surveillance Tech to Fight New Coronavirus Spread.

Maghazei, O., Netland, T., 2020. Drones in manufacturing: exploring opportunities for research and practice. J. Manuf. Technol. Manag. 31, 1237−1259. Available from: https://doi.org/10.1108/JMTM-03-2019-0099.

Mammeri, A., Siddiqui, A.J., Zhao, Y., 2021 UAV-assisted railway track segmentation based on convolutional neural networks. In: Proc. of the 2021 IEEE 93rd Vehicular Technology Conference (VTC2021-Spring), pp. 1−7.

Moses-Gonzales, N., Brewer, M.J., 2021. A special collection: drones to improve insect pest management. J. Econ. Entomol. 114, 1853−1856. Available from: https://doi.org/10.1093/jee/toab081.

Moshou, D., Bravo, C., Oberti, R., West, J.S., Ramon, H., Vougioukas, S., et al., 2011. Intelligent multi-sensor system for the detection and treatment of fungal diseases in arable crops. Biosyst. Eng. 108, 311−321. Available from: https://doi.org/10.1016/j.biosystemseng.2011.01.003.

Mulla, D.J., 2013. Twenty five years of remote sensing in precision agriculture: key advances and remaining knowledge gaps. Biosyst. Eng. 114, 358−371. Available from: https://doi.org/10.1016/j.biosystemseng.2012.08.009.

Najiya, K.V., Archana, M., 2018 UAV video processing for traffic surveillence with enhanced vehicle detection. In: Proc. of the 2018 Second International Conference on Inventive Communication and Computational Technologies (ICICCT), pp. 662–668.

Naqvi, S.A.R., Hassan, S.A., Pervaiz, H., Ni, Q., 2018. Drone-aided communication as a key enabler for 5G and resilient public safety networks. IEEE Commun. Mag. 56, 36–42. Available from: https://doi.org/10.1109/MCOM.2017.1700451.

Nguyen, M.T., Truong, L.H., Le, T.T.H., 2021. Video surveillance processing algorithms utilizing artificial intelligent (AI) for unmanned autonomous vehicles (UAVs). MethodsX 8, 101472. Available from: https://doi.org/10.1016/j.mex.2021.101472.

Niccolai, A., Grimaccia, F., Leva, S., 2019. Advanced asset management tools in photovoltaic plant monitoring: UAV-based digital mapping. Energies 12. Available from: https://doi.org/10.3390/en12244736.

Nooralishahi, P., Ibarra-Castanedo, C., Deane, S., López, F., Pant, S., Genest, M., et al., 2021. Drone-based non-destructive inspection of industrial sites: a review and case studies. Drones 5. Available from: https://doi.org/10.3390/drones5040106.

Orlowski, C.T., Girard, A.R., 2012. Dynamics, stability, and control analyses of flapping wing micro-air vehicles. Prog. Aerosp. Sci. 51, 18–30. Available from: https://doi.org/10.1016/j.paerosci.2012.01.001.

Panday, U.S., Pratihast, A.K., Aryal, J., Kayastha, R.B., 2020. A review on drone-based data solutions for cereal crops. Drones 4.

Pantazi, X.E., Moshou, D., Bochtis, D., 2019 Intelligent Data Mining and Fusion Systems in Agriculture. ISBN 9780128143926.

Radoglou-Grammatikis, P., Sarigiannidis, P., Lagkas, T., Moscholios, I., 2020. A compilation of UAV applications for precision agriculture. Comput. Netw. 172, 107148. Available from: https://doi.org/10.1016/j.comnet.2020.107148.

Rahman, M.F.F., Fan, S., Zhang, Y., Chen, L., 2021. A comparative study on application of unmanned aerial vehicle systems in agriculture. Agriculture 11. Available from: https://doi.org/10.3390/agriculture11010022.

Ramezani, A., Chung, S.-J., Hutchinson, S., 2017. A biomimetic robotic platform to study flight specializations of bats. Sci. Robot. 2. Available from: https://doi.org/10.1126/scirobotics.aal2505.

Restás, Á., 2022. Drone applications fighting COVID-19 pandemic—towards good practices. Drones 6. Available from: https://doi.org/10.3390/drones6010015.

Restás, Á., Szalkai, I., Óvári, G., 2021. Drone application for spraying disinfection liquid fighting against the COVID-19 pandemic—examining drone-related parameters influencing effectiveness. Drones 5. Available from: https://doi.org/10.3390/drones5030058.

Reyes, I.O., Beling, P.A., Horowitz, B.M., 2017. Adaptive multiscale optimization: concept and case study on simulated UAV surveillance operations. IEEE Syst. J. 11, 1947–1958. Available from: https://doi.org/10.1109/JSYST.2015.2503395.

Saeed, F., Mehmood, A., Majeed, M.F., Maple, C., Saeed, K., Khattak, M.K., et al., 2021. Smart delivery and retrieval of swab collection kit for COVID-19 test using autonomous unmanned aerial vehicles. Phys. Commun. 48, 101373. Available from: https://doi.org/10.1016/j.phycom.2021.101373.

Shafiee, M., Zhou, Z., Mei, L., Dinmohammadi, F., Karama, J., Flynn, D., 2021. Unmanned aerial drones for inspection of offshore wind turbines: a mission-critical failure analysis. Robotics 10. Available from: https://doi.org/10.3390/robotics10010026.

Sharkey, N., 2011. The automation and proliferation of military drones and the protection of civilians. Law Innov. Technol. 3, 229–240. Available from: https://doi.org/10.5235/175799611798204914.

Smith, M., Pars is a Search and Rescue Drone Capable of Saving Lives.

Stolaroff, J.K., Samaras, C., O'Neill, E.R., Lubers, A., Mitchell, A.S., Ceperley, D., 2018. Energy use and life cycle greenhouse gas emissions of drones for commercial package delivery. Nat. Commun. 9, 409. Available from: https://doi.org/10.1038/s41467-017-02411-5.

Tagarakis, A.C., Filippou, E., Kalaitzidis, D., Benos, L., Busato, P., Bochtis, D., et al., 2022. For 3D mapping of orchard environments. Sensors 22. Available from: https://doi.org/10.3390/s22041571.

Thomazella, R., Castanho, J.E., Dotto, F.R.L., Júnior, O.P.R., Rosa, G.H., Marana, A.N., et al., 2018. Environmental monitoring using drone images and convolutional neural networks. In: Proc. of the IGARSS 2018 − 2018 IEEE International Geoscience and Remote Sensing Symposium, pp. 8941−8944.

Velusamy, P., Rajendran, S., Mahendran, R.K., Naseer, S., Shafiq, M., Choi, J.-G., 2022. Unmanned aerial vehicles (UAV) in precision agriculture: applications and challenges. Energies 15. Available from: https://doi.org/10.3390/en15010217.

Wang, X., Chowdhery, A., Chiang, M., 2017. Networked drone cameras for sports streaming. In: Proc. of the 2017 IEEE 37th International Conference on Distributed Computing Systems (ICDCS), pp. 308−318.

Watts, A.C., Ambrosia, V.G., Hinkley, E.A., 2012. Unmanned aircraft systems in remote sensing and scientific research: classification and considerations of use. Remote Sens. 4.

Weibel, R.E., Hansman, R.J., 2004 Safety considerations for operation of different classes of UAVs in the NAS. In: Proc. of the 4th Aviation Technology, Integration and Operations Forum, AIAA 3rd Unmanned Unlimited Technical Conference, Workshop and Exhibit, Chicago, Illinois.

Wood, R.J., 2007 Liftoff of a 60mg flapping-wing MAV. In: Proc. of the 2007 IEEE/RSJ International Conference on Intelligent Robots and Systems, pp. 1889−1894.

Yun, C.Y., Park, I., Lee, H.Y., Jung, J.S., Hwang, I.S., 2004 A new VTOL UAV cyclocopter with cycloidal blades system. In: Proc. of the American Helicopter Society 60th Annual Forum, Baltimore.

Zachariadis, O., Mygdalis, V., Mademlis, I., Nikolaidis, N., Pitas, I., 2017 2D visual tracking for sports UAV cinematography applications. In: Proc. of the 2017 IEEE Global Conference on Signal and Information Processing (GlobalSIP), pp. 36−40.

Zhang, J., Campbell, J.F., Sweeney II, D.C., Hupman, A.C., 2021. Energy consumption models for delivery drones: a comparison and assessment. Transp. Res. D Transp. Environ. 90, 102668. Available from: https://doi.org/10.1016/j.trd.2020.102668.

CHAPTER 2

A drone view for agriculture

Bojana Ivošević[1], Marko Kostić[2], Nataša Ljubičić[1], Željana Grbović[1] and Marko Panić[1]

[1]BioSense Institute, University of Novi Sad, Novi Sad, Serbia
[2]Faculty of Agriculture, University of Novi Sad, Novi Sad, Serbia

Introduction

This chapter covers an introduction, to the types of unmanned aerial systems (UASs) and their onboard sensors, mission-planning, and state-of-the-art technology used in UASs for agricultural applications. High-resolution images acquired from UAS offer detailed information about crop and soil condition being useful for agriculture management purposes. Knowledge and understanding about usage and limitations of UAS data and related software tools are necessary as their market grows.

Why does agriculture need a drone view?

Modern concepts in agriculture force engagement of sophisticated, fully automated, and massive mobile systems, which offer advances in terms of in-field productivity, spatially related treatment and therefore input savings, payback of investments, and profit per unit of field area. Furthermore, simplification of the natural environment is the way in which humankind strives to ensure efficient exploitation of resources (Addicott, 2020) through specialization of agricultural holdings.

In other words, the trend of introduction of hyperproductive agricultural mechanization is continuous and imposed by the need for resizing production units (fields) in terms of increasing their size by merging and grouping smaller ones into unified large agricultural holdings. As the size of farms increases, new issues have emerged. Field merging contributes to the increase of in-field heterogeneity mainly because the initial field units had been exposed to decades of specific treatments by their previous farm owners, or were initially separated into independent fields respecting environmental distinction such as relief structure, channel routes, and hedges (Oliver, 2010).

Unmanned Aerial Systems in Agriculture
DOI: https://doi.org/10.1016/B978-0-323-91940-1.00002-5

In the modern agriculture era, the role of the farmer has changed, dedicating less time to weather-dependent field work. The number of workers in farming industries and number of people inhabiting countryside areas has reduced dramatically, so the human time engagement per area unit of a field has been in a continuous decline.

Throughout history, visual scouts have been critical for the success of agriculture. Farming transformation has led to the weakening of intuitive relationships between the farmer and his land, which has restricted a detailed insight into the temporal process (Fig. 2.1). It is estimated that in 1880, 20 human hours were needed to harvest one acre, while in 2015 only 3.5 hours of human labor were spent for harvesting a field sized 6.32 ha (Addicott, 2020).

Scouts continue to be the key to success; only the sophistication of the tools used for scouting operations has changed. UASs are becoming an increasingly applied technology that provides unique insight into the entire field area at a glance. This technology has the potential to give farmers a whole new perspective on their operations—from above. Even though, at a global extent, there are large differences in agricultural practices influenced by the cultural, sociological, and financial background that attenuate wide acceptance of UAS technology (Mogili and Deepak, 2018).

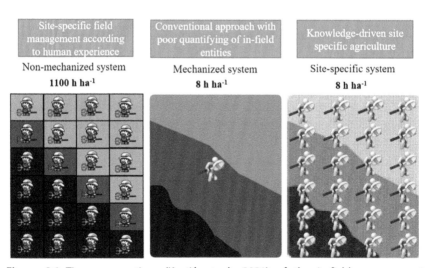

Figure 2.1 Time perspectives (Kostić et al., 2021) of the in-field management approach (Ibarrola-Rivas et al., 2016). *Data on time consumption originated from Ibarrola-Rivas, M., Kastner, T., Nonhebel, S., 2016. How much time does a farmer spend to produce my food? An international comparison of the impact of diets and mechanization. Resources 5, 47. https://doi.org/10.3390/resources5040047.*

Western countries are advancing in the application of UASs in the framework of precision agriculture (Zhang and Kovacs, 2012), using photogrammetry, and remote sensing. There are obvious benefits from using UAS technology regarding information gathering in the precision agriculture practice, but some concerns must be highlighted such as the lack of a robust decision support system for automatic recognition of spatial patterns and its temporal variation (McBratney et al., 2005).

As the world population continues to grow, we need innovative ways to sustainably feed more people. This means leveraging intelligence in agriculture to make operations smarter and more efficient. Managing crops is a labor-intensive, tedious, and time-consuming process. Especially in developing countries, like Serbia, agricultural practices rely on a farmer's traditional knowledge with no scientific background, often resulting in inadequate use of fertilizers, low productivity, and crop damage.

In the 21st century, agricultural robots are at the center of the smart agriculture expansion, among which, UASs have gained most of the attention (Avtar and Watanabe, 2020; Milics, 2019; Mogili and Deepak, 2018; Zhang and Kovacs, 2012). UAS technology comes from a military background where it is used for reconnaissance purposes such as territorial monitoring and direct combat purposes. However, the recent proliferation of nonmilitary applications of UASs has spread into many industries (Fig. 2.2).

To be highly capable of efficiently fulfilling tasks in precision agriculture (such as sensing and scouting, fertilization management, pest management,

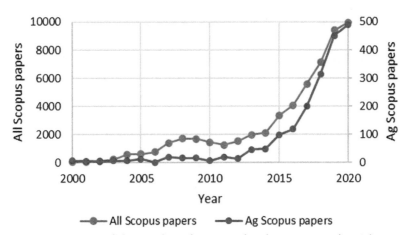

Figure 2.2 Summary of the number of papers related to unmanned aerial systems published in scientific journals noted by Scopus from 2000 to 2020.

plant counting, biomass prediction, yield prediction, livestock management, and management zones delineation), agricultural UAS can be utilized combining information and communication technologies, artificial intelligence (AI), big data, and internet of things (IOTs) (Avtar and Watanabe, 2020; Urbahs and Jonaite, 2013). In agriculture, UASs are flying platforms most frequently engaged in aerial auditing with special cameras, but recently the role has widened to spraying applications or even seeding, especially on inaccessible areas (Milics, 2019).

Timing plays a crucial role in the world of agriculture. When crops are in danger, every minute counts. Plant stress caused by pests, water deficiencies, or lack of nutrients sometimes requires decision-making on the spot. That is where the UAS as an agricultural toolkit proves its efficacy by bringing actionable data of plant health in a real time.

The main advantage using an unmanned aircraft compared to other devices used in remote sensing (satellites, piloted aircrafts, etc.) is the opportunity to attain high temporal and spatial resolution. The temporal resolution is reflected in the flexibility of the flight time, which is very important when considering the duration of certain vegetative phases and their conditionality from the applied agro-technics and agroecological conditions. Spatial resolution involves high-resolution records of as much as a few centimeters per pixel and depends on the type of sensor and flight height during recording. With a decrease in flight altitude, the resolution of records increases, but the coverage of the area of interest decreases. Equipped with adequate sensors, UASs deliver data at a fine spatial scale whereas temporal resolution is defined by the end user. Satellites, however, are covering large areas, but provide coarse spatial resolution and images are very often influenced by cloud and fog cover. Although some of the commercial satellites can provide subdecimeter resolutions, the major drawback is that they are infrequent in time (Avtar and Watanabe, 2020).

The majority of commercially available UASs are sensitive to weather conditions, especially wind disturbances, heavy rains, and sudden changes in sun illumination when considering data collection consistency. UAS flight itself can be affected by adverse weather and have potential implications on data collection. However, the impact of atmospheric elements on the quality of UAS footage is smaller compared to satellites or planes, because the sensor is closer to the surface it records. UASs are positioned below the clouds, somewhere between satellites and ground vehicles, thus bridging the gap between satellite and high-altitude aerial platforms on one hand, and ground-based observations on the other.

From an agroecological point of view, acquiring data with an UAS is a noninvasive method, offering farmers detailed insight of crops from low altitudes and leaving the crops untrodden. Moreover, equipped with rechargeable batteries, they are safe for the environment and create win-win situations for agriculture and biodiversity (Ivosevic et al., 2015; Librán-Embid et al., 2020). UAS allows for acquiring and storing a large posse of data and time series of crops. This technology provides significant economic advantages by reducing human labor, working hours, and the number of chemical inputs. UASs have been developed into affordable, earth-and-user-friendly sophisticated flying machines.

The shortcomings of UAS technology are reflected in limited possibilities when it comes to collecting information from larger areas. Commercially available UASs are usually smaller in size and are powered by a rechargeable battery that has limited capacity. Some of the well-known technical problems of a small UAS are engine power, short flight duration, difficulties in maintaining the flight altitude, and aircraft stability due to wind and turbulence (Zhang and Kovacs, 2012).

Another key issue is the sensor selection restriction by cause of the UAS payload weight. Generally, a payload weight for an UAS is 20%−30% of the total system weight, which poses limitations to the sensing systems that can be mounted. Attached sensors are generally much more expensive than the aircraft itself, and their cost can be intimidating for most farmers and practitioners. Moreover, proper handling of a UAS requires certain knowledge and skills. This does not only mean flying skills, but also knowledge of platforms and sensors to successfully carry out tasks in agriculture.

Storing and maintaining the equipment in the right way is very important for precision and longevity. Issues of UAS image processing and application are generally similar to those for traditional aerial and satellite images. They include instrument calibration, atmospheric correction, vignetting correction, line-shift correction, band-to-band registration, and frame mosaicking (Zhang and Kovacs, 2012).

Furthermore, data processing and manipulation itself requires knowledge in the field of technology and handling of certain software that allows visualization and interpretation of UAS images, maps, and other results. Drones are a relatively new technology that is not mature yet, and therefore lack of regulation can cause issues. Inadequate handling and noncompliance of regulations pose a threat to the general public, property, and the environment.

It is important to point out that, without domain knowledge, the very existence of this technology is worthless if it is not available to farmers. Even backed up by proper research, many of the agricultural tasks still could not be fulfilled by UAS service providers (Veroustraete, 2015).

When it comes to data collection, it would involve careful mission-planning according to the specifications of the field, including preparation of the flight itself, comprehensive knowledge of the sensors that collect data, optionally real-time kinematic (RTK) corrections, postflight checking, data processing, and finally analyzing data and results interpretation.

Types of unmanned aerial systems

UASs can be classified by a wide range of performance characteristics. When considering their external appearance and design for agricultural purposes, there is a common distinction between copter, fixed wing, and vertical takeoff and landing (VTOL) systems. Each UAS type has its advantages and disadvantages concerning area coverage, flight time, ease of use, payload capabilities, requirements for takeoff and landing, availability on the market, price, etc. (Milics, 2019; Ramesh and Muruga, 2020). Table 2.1 shows some of the main characteristics of all three mentioned types of UASs.

The most popular type of consumer-grade UAS is a multirotor or multicopter type, whose name implies that it consists of numerous rotors that generate lift required for flight. The most common multirotor type of drone has four rotors; quadcopter. Hexacopters (with six rotors) and octocopters (with eight rotors) are slightly less prominent and are usually used for their stability and ability to carry heavy payloads.

The main advantage of this type of aircraft is their maneuverability and ease of use. Multirotor drones are used when high-precision surveillance is needed because of their ability to slow down and hover. Not only can they hover steadily in one place, but they can also maintain low altitudes such as a few meters from the ground when needed. They can fly in narrow spaces and around obstacles. They constantly consume loads of power from the battery to spin the propellers enough to generate lift. Moreover, they can take off and land vertically from any location.

Their disadvantage is less flight autonomy due to higher energy consumption, and therefore less terrain coverage. They can also face technical limitations due to communication distance because they have smaller range than other types of drones. Despite these limitations, they are

Table 2.1 Unmanned aerial system types (UAS).

Type of UAS		Pros	Cons
Multirotor		• Accessibility • Ease of use • Hovering • Small take-off and landing area • Can support heavy payloads • Flies in low altitudes • Affordable price	• Short flight time • Small area coverage • Small wind resistance
Fixed wing		• Covers large areas • Large endurance • High flying speed	• Launch and recovery requires a lot of space • No VTOL and hovering • More difficult to fly • Expensive • Supports lighter payload
VTOL		• Vertical take-off and landing • Large area coverage • Hovering mode • Supports heavier payload • Intermediate or no pilot experience • High wind resistance	• Expensive

valuable tools in precision agriculture for reconnaissance and mapping smaller fields of around 8 ha. They can carry payloads such as red—green—blue (RGB), multispectral, hyperspectral, or thermal cameras to collect data from the field. It is not unusual for these drones to carry multiple sensors at the same time. The number of rotors corresponds to the payload and UAS size differences (Kim et al., 2019). Rotary types, especially the ones with a higher number of rotors, have greater payload capabilities. They are widely used for variable rate application of liquid pesticides or fertilizers; especially the hexacopter or octocopter types can carry tanks weighing more than 10 kg. Furthermore, multirotor UASs are used for tasks that require extreme precision, such as plant counting, pollen moisture distribution, or even sowing.

A fixed-wing aircraft is a type of drone that has a specific aerodynamic shape that resembles a manned airplane (Kim et al., 2019). Fixed-wing unmanned aerial vehicles (UAVs) require a catapult or hand-launch to takeoff and a parachute or soft landing space but require less energy to carry a unit weight of cargo. However, they are unable to carry heavy payloads because there is a high possibility of damage to the belly area when landing.

They are suitable for covering larger areas but with less precision, which means lower spatial resolution compared to rotary types. Fix wings are fitted to larger projects covering approximately a range of 500—750 acres per hour and reduce required flight time flying at a higher speed ranging from 25 to 45 mph (Puri et al., 2017). They cannot float or hover because to maintain their height, they must move constantly. Due to this feature, they are specialized to fly at higher altitudes. Compared to the multirotor type, the fixed-wing drone needs a pilot with a certain level of confidence and competence in flying skills to control it from launch, through the flight, and then bring it back to a soft landing. Other downsides are their higher cost compared to rotary aircrafts. A fixed-wing UAS in agriculture is mainly used in large areas, for scouting and monitoring fields, spraying, detecting pests, etc.

VTOL is a type of UAS that combines features of both multirotor and fixed-wing aircraft. It is considered as a hybrid aircraft and has not yet reached its full potential of commercialization. These drones are adapted to cover larger areas and can take off, hover, and land vertically using propellers located at the front and/or rear of the aircraft. They take off vertically, and when they reach the appropriate height, they are placed at a horizontal flight position. They also have the aptitude to tilt to a certain level to compensate for the wind, ensuring that the aircraft maximizes stability and safety during landing and takeoff. VTOL systems bring together the precision of the multirotor takeoff and landing with the coverage of a fixed-wing aircraft. Since the payload weight and quality limits increase, their design allows for higher accuracy than fixed wings. Depending on the altitude, they can cover approximately 1000–1800 acres (around 400–700 ha).

VTOL systems are adapted to fly in toughest conditions (windy weather). Their flight distance is exceptionally large, some even completing their tasks beyond communication range. As their payload capacity is bigger, VTOL systems can carry heavier cameras with higher resolutions, light detection and ranging (LIDAR), or other sensors, since they offer a safe and controlled landing. This type of UAS is fully autonomous and does not require runways or advanced piloting skills. In the world of unmanned aerial technology, VTOL are new systems which are more expensive than both fixed wing and multirotor aircrafts. VTOL systems are suitable for agriculture applications where covering large areas is necessary, but also where sophisticated and professional surveying and mapping is required. They are applicable for all tasks from monitoring crops and detecting anomalies in the field, to biomass estimation and yield prediction.

Mission-planning

The data collection phase begins with the flight mission planning, marking geospatial checkpoints for proper positioning (GCPs), and the UAS flight itself. The drone control system can be semiautomatic or automatic. With the semiautomatic control mode, it is necessary to have a professionally trained person to enter the flight parameters, as well as take control of the aircraft in cases of unsafe terrain (usually when landing, avoiding obstacles, and sudden adverse weather conditions). The person creating the mission can manipulate parameters such as: terrain dimensions, flight altitude, the degree of overlap between adjacent images (front-lap and side-lap), camera angle and features related to image characteristics. Fully autonomous drones have control that allows them to perform all predefined tasks related to flight, flight path, shooting and landing, automatically without the assistance of the operator (Wich and Koh, 2018).

Newer drones can recognize obstacles, such as power lines, buildings or trees and have the option of automatic avoidance. Collecting data in a precise way with a clearly set goal is essential for efficiently diagnosing of differences within crops. A key aspect of all UAS missions for acquisition of accurate data is setting the ideal combination of sensor parameters and flying height to yield images of sufficient resolution. These settings depend on the type of the task that needs to be accomplished such as plant counting, biomass estimation, detecting anomalies of the crops, water or nutrient deficiencies, or pest infestation. Obviously, small objects will need a higher ground resolution that can be achieved with flying at lower altitudes. Nevertheless, flying lower is not necessarily appropriate for all the potential issues that should be addressed. It will depend on lots of factors such as vegetation height and growth, the complexity of the terrain, and capabilities of the sensors and drone.

Besides flying altitude, the flight mission requires aircraft stability to prevent errors and blurring that results from differences in brightness and image destabilization (Ivosevic et al., 2017). Flying a UAS at high speed will most probably capture blurred images due to slow shutter speed. Furthermore, intense winds can significantly affect the motion of the drone resulting in unsharp images.

Ground resolution often called ground sampling distance (GSD) can be calculated automatically through ground control software and represents the distance between the center of two adjacent pixels in the image. It can be calculated without a software, if the correction factors such as focal length and pixel size for the camera is known, using the following equation:

$$GSD = Pixelsize * \left(\frac{AGL}{Focallenght} \right) \tag{2.1}$$

Pixel size can be obtained from sensor resolution and sensor size. For example, if sensor resolution is $X \times Y$ and the sensor size is A (mm) $\times$ B (mm), dividing either A by X or B by Y results the pixel size in mm. AGL is the aboveground level in meters. Focal length should be known from the manufacturer or user's manual (Wich and Koh, 2018).

The drone pilot requires a certain experience in the field and knowledge of the law on drone regulations to be always up to the task. This primarily refers to flying capabilities and composure for the task to be performed accurately, and the safety of third parties ensured.

Setting a proper flight mission requires a carefully designed flight path covering the boundaries of the region of interest (ROI). This path is a route or trajectory which an UAS will follow once it takes off, taking pictures automatically on the way. Creating a flight plan relies on capabilities of the sensors, for example, resolution and focal length, and mission-planning software performance. UAS mission paths can have different shapes such as rectangular, circular, or arbitrary.

Usually for mapping a crop field, 2D maps are generated. For detailed projects and when creating 3D models, double flight grid mission is required meaning the path will be crossed. One flight has a trajectory of image acquisition parallel while the other has an orthogonal position to the given object. This setting aims to take sufficient images from different perspectives to create a proper 3D model. Depending on the nature of the problem that needs to be addressed, the camera can be set to collect images from a nadir perspective (90-degree angle or perpendicular to the ground) or at some specific angle when oblique images are needed. When the objective is to collect information of vegetation indices, nadir images are the only choice.

Camera parameters should also be adapted to weather and environmental conditions such as sun illumination. For acquiring data to extract vegetation indices, some systems require camera settings in a manual mode. This implies customized settings of ISO (Sensitivity of the camera's sensor), white balance, shutter speed, and other camera setting parameters.

Smart agricultural practices require precision and with UASs it is possible to acquire spatial resolution and precision at centimeter level via setting ground control points (GCPs) and using RTK or postprocessing kinematic (PPK). The corrected GNSS (Global navigation satellite system) path accuracy will depend on performances of the device used, current satellite availability, and receipt quality. RTK modifies coordinate positions (latitude, longitude, and altitude) in real time and rewrites them into EXIF (Exchangeable image file format) data of each image while the UAS is still in the flight mission.

A mobile station provides real-time corrections at centimeter-level accuracy ensuring that each photo uses the most accurate metadata. Another option is PPK where the coordinate corrections are done after the flight mission and EXIF coordinates are written into jpg or tiff files. These data are used in the preprocessing step before the creation of the orthomosaic or other photogrammetry outputs. GCPs are marked points on the ground that have a known geographic location. GCPs are required in aerial imaging because they enhance the positioning and accuracy of the mapping outputs and reduce the noise. They should be placed on the ground and be clearly visible in the aerial imagery by using high contrast colors. It is also important that the GCPs are evenly distributed over the whole mapping area. These coordinates can be captured via ground-based RTK GPS or obtained by other sources such as LIDAR, older maps of the area, and web map services. After downloading, the acquired UAS images can be processed in the photogrammetry software such as Pix4Dmapper (2021 Pix4D SA, Switzerland). The software includes three main steps:

1. Initial processing.
2. Point cloud and mash.
3. Produce digital surface model (DSM), orthomosaics.

Known GCPs should be included in initial processing step before creating the final orthomosaic.

Sensor availability

The basic function of UASs in the agriculture niche is to collect high-quality images. Accordingly, sensors that could be mounted on an UAS should be chosen carefully depending on various tasks that need to be accomplished. Sensors can be divided according to the method by which they achieve spatial and spectral discrimination (Boreman, 2005) ranging from visible light sensors (RGB), infrared (IR) or thermal sensors, multispectral sensors, hyperspectral to LIDAR. The first five listed types of sensors are most utilized (Fig. 2.3), while laser detectors are well applied in environmental sciences for terrestrial scanning. This technology known as LIDAR became available recently and till now it is rarely exploited in agricultural studies due to the increased cost.

Furthermore, an ongoing miniaturization of measurement technologies, that is, smaller payloads and developments in navigational capabilities

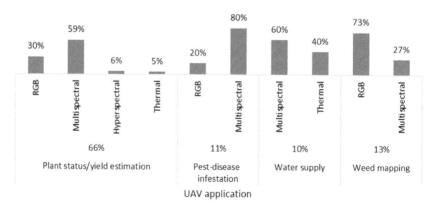

Figure 2.3 Share of unmanned aerial system application in specific tasks in agriculture and sensing platforms utilized in various aspects.

will enable datasets that, until now, farmers have only dreamed of collecting. However, the needs for low payload weight and the engagement of small flying platforms can impose several limitations on the selections of the sensors to be used.

An important consideration when using UASs to collect data is to determine whether the system will need one or multiple sensors, or if data can be collected over multiple flights. Since recently, it is a widespread practice to combine two or more sensors to gain a broader perspective of information from different ranges of wavelengths or even different types of data. This is particularly important when it comes to a potential problem or anomaly in the field which is unexplored and yet to be discovered and spotted in a specific wavelength range. Collected information from sensors in a form of a range of wavelengths that represents a distinctive spectral signature of a plant is a valuable asset for farmers in tasks where there is an urge for immediate decision-making.

Visible light sensor (red—green—blue camera)

Image-based approaches to measure plant morphological traits rapidly and nondestructively have emerged and quickly developed in response to the need for accelerating high-throughput plant phenotyping that would enable effective use of genomic data to bridge the genotype-to-phenotype gap for crop improvement. RGB cameras are mostly common sensors attached to UAS for precision agriculture applications. This sensor captures the red, green, and blue channels of visible light which provides the image color to the human eye.

Through the affordability of high-resolution digital cameras, RGB UAS remote sensing with centimeter resolution can easily be obtained. Barbosa et al. (2021) evaluated the potential of the application of UAS and RGB vegetation indices in the monitoring of a coffee crop. They evaluated nine Vis (vegetation indices) in this study using a DJI Phantom 3 professional UAS (DJI, Shenzhen, China) equipped with a digital RGB camera (Sony brand, model EXMOR $\frac{1}{2}0.3''$) with a resolution of 4000×3000 pixels, sensor of $6.16 \text{ mm} \times 6.62 \text{ mm}$, field of view of 94 degrees, and a sampling rate 2 Hz. The flying altitude was set to 30 m and speed 3 ms^{-1} with 80% overlap between images. From the RGB spectrum, several Vis were calculated selecting certain equations addressing diverse sources. In their general conclusion, they highlighted that RGB is not robust enough for reliable leaf area index (LAI) estimates in the coffee production. Great advantage of RGB sensing is the fact that the information acquired requires simple processing. Flight missions with imaging can be executed in different environmental conditions, but a specific time frame is required based on weather conditions to avoid inadequate or excessive exposure of the image (Tsouros et al., 2019).

The main weakness of the RGB sensing from the drone is the limited possibilities for distinction of crop or soil characteristics due to poor spectral information. RGB cameras are often fused with other type of sensors. In the study of Wan et al. (2020), the normalized difference yellowness index (NDYI) drawn from RGB spectrum was the most useful index to track the changes in leaf chlorophyll content as well as the leaf greenness during the whole growth period, while fusion of the multitemporal normalized difference vegetation index (NDVI) canopy height and canopy coverage achieved the best prediction of grain yield with a determination coefficient of 0.85 and 0.83.

Infrared or thermal sensor

Thermal imaging systems use mid- or long wavelength IR energy sensing differences in heat. Thermal remote sensing has immense potential for use in agriculture, but they are not as common in UASs as other sensors, such as optical and multispectral sensors, and this is likely owing to the features of thermal sensors and the data that can be gathered from them.

Thermal energy is detected by converting absorbed EM (Electromagnetic) radiation into quantified values (Messina and Modica, 2020). Thermal imagery is specified by a lower spatial resolution and contrast and by a stronger optical distortion (Boesch, 2017). Cost affordable thermal cameras can only provide data about relative temperature differences, while using radiometrically calibrated UAS cameras provide accurate surface temperature. If high-altitude measurements are taken, air circumstances have an impact on the quality of the thermal picture. Only by taking measurements within 10 m of the target's surface can the effects of atmospheric conditions be avoided. Also, crop water stress index (CWSI) was shown to function better in dry climates, whereas it revealed significant limits in wet climates and areas with high climatic fluctuation.

The temperature of crop canopy surface is in relation to the transpiration rate and indirectly to crop available soil water status. If plants suffer from water deficiency, stomata limits transpiration and evaporative cooling, which results in an elevated leaf temperature, so thermal imaging can be used to identify water stress conditions in plants. Also, ambient and soil temperature have crucial roles in metabolic processes in the plants and bio-chemical soil processes. Measurements of stem water potential and stomatal conductance, the most prominent physiological water stress indicators used to estimate agricultural water status, are required when using UASs in the research of plant water status. Thermal images showed better sensitivity to water stress that are undetectable by the NDVI (Baluja et al., 2012).

Some research was conducted primarily to predict availability of soil moisture, plant-based temperature, and evapotranspiration using UAS-thermal imaging, and subsequently to optimize schedule of irrigation. For example, Lacerda et al. (2022) used UAS-based images collected from cotton fields in the thermal infrared waveband to calculate CWSI and evaluate them as predictors of leaf water potential (LWP). Data were collected using a 3Dr Solo quadcopter (3D Robotics, Berkeley, CA, the United States) equipped with a FLIR Vue Pro R (Model 640, 69 degrees FOV, 9 mm, 30 Hz; FLIR Systems, Inc., Wilsonville, OR, the United States) camera that collects 14-bit images in the 7.5−13.5 μm region of the electromagnetic spectrum. The authors stated that the obtained results showed the potential of using an affordable UAS-thermal system to produce predicted LWP maps that are representative of the current field water status in the humid southeastern USA. Crusiol et al. (2020) evaluated the water status of soybean plants under various water circumstances.

The utilization of produced maps can help farmers in better monitoring of crop water status spatial variability during the season and assist in improving irrigation management decisions. Generated maps can be potentially utilized to design temporally dynamic variable rate irrigation systems based on crop and type of irrigation management.

Multispectral

The multispectral sensors can detect RGB bands, plus optionally near-IR, red edge, or thermal bands. A higher number of radiometric bands offer an extended number of calculations, thus more vegetation indices. Multispectral sensors have advanced performance in plant phenotyping due to increased number of vegetation indices that can be derived. In contrast to hyperspectral sensors, the procedure for multispectral sensor calibration is consistent and typically involves using a known reference object or target to measure the sensor's response to different wavelengths of light (Esposito et al., 2021). The goal of radiometric calibration is to ensure that the sensor is measuring the true radiance or reflectance values of the target object and is not affected by any errors or biases in the sensor's hardware or software. Radiometric calibration is an essential step in ensuring the accuracy and reliability of multispectral sensor measurements.

Downside of multispectral sensors, apart from the higher purchase price, is the lower output image resolution compared to RGB images which impose lower flight altitude to obtain an adequate ground resolution for the surveyed objects and to avoid missing data. In the study of Furukawa et al. (2021), multispectral UAS, as compared to the RGB UAS, had lower overall accuracies, although the vegetation class of the RGB UAS presented high producer's and user's accuracy over time, comparable to the multispectral UAS results. The authors inferred that the multispectral UAS is more suitable in characterizing vegetation, bare soil, and dead matter classes on landslide areas while the RGB UAS can deliver reliable information for vegetation monitoring.

Multispectral imaging may be more helpful for answering research questions such as the interaction between genotype, environment, and management because it allows for better resolution of the relationship between plant physiological status and sensor data, which could be used to improve understanding of yield gaps (Herzig et al., 2021). In a relevant study of Herzig et al. (2021), the aim was to evaluate performance of RGB and multispectral UAS imagery in high-throughput phenotyping and yield prediction in barley breeding. The yield prediction accuracy of multispectral systems was at the same level as RGB and reached a maximum prediction accuracy of $R^2 = 0.82$. Overall conclusion was that due to the lower costs and the consumer-friendly handling of image acquisition and processing, the RGB imagery seems to be more suitable for yield prediction in this study.

Hyperspectral

UASs observations by using either RGB or NIR (Near infrared) sensors are the most frequent setup of the trials in agricultural applications. Still, it lacks in terms of spectral range for detailed insight of profile materials and organisms that only multispectral and hyperspectral sensors (Fig. 2.4) can ensure. Formerly, the multi/hyperspectral land observation technology was initially available only for orbital platforms and afterwards manned aircraft for monitoring vegetation and environmental parameters (Pádua et al., 2017).

Figure 2.4 Hyperspectral camera (Zolix instruments CO., Ltd, Beijing, China) carried by an unmanned aerial system.

Advancement in sensing technology reduced the size of HS/MS (Hyperspectral/Multispectral) cameras and made them available for commercial use that can be attached to the UASs for scientific or public purposes. There is immense potential of UAS application in agriculture to assess plants' properties such as water status (Zhao et al., 2020), biomass and yield estimation (Roy Choudhury et al., 2021), and disease monitoring (Mahlein et al., 2017) is widely recognized. Hyperspectral imagery contains a high number of spectral bands (hundreds or thousands) arranged in a narrow bandwidth from 5 to 20 nm each.

Hyperspectral data allow spectral composition of narrowbands known as spectral signature unlike multispectral data that include wider spans of the EM spectrum providing utmost possibilities in materials characterization. Hyperspectral imaging is a chemical-free assessment method that requires minimal sample preparation. Therefore it saves labor, time, and reagent cost (Wu and Sun, 2013). Coarser spectral resolution of multispectral data cannot recognize specific spectral features which restrict its ability in plant and soil detection, and could shade certain relevant components. Powering sensing capabilities in terms of spectral bands implicate harder data processing especially if on-the-go decisions are required.

So multiplying or splitting the bands to narrower ones requires more computational resources. As a passive detector, HS records are in most cases loaded by different impacts of ambient conditions such as light

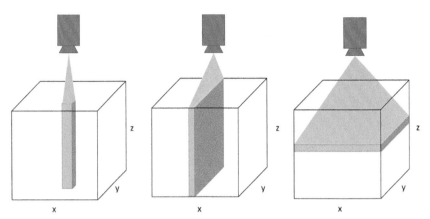

Figure 2.5 3D hyperspectral datacube and acquisition approaches of hyperspectral images; point scanning (left), line scanning (middle), plan scanning (right). *Adapted from Wu, D., Sun, D.-W., 2013. Advanced applications of hyperspectral imaging technology for food quality and safety analysis and assessment: a review—part I: fundamentals. Innov. Food Sci. Emerg. Technol. 19, 1–14. https://doi.org/10.1016/j.ifset.2013.04.014.*

intensity and exposure, and defined protocols need to be conducted to accurately connect spectral data with an object feature. Taking into account acquisition modes, hyperspectral imaging can be designated in three separate categories (Wu and Sun, 2013): point scanning, line scanning, and plan scanning (Fig. 2.5).

There are various applications of hyperspectral cameras carried by UASs in agriculture. Zarco-Tejada et al. (2013) used high-resolution hyperspectral imagery acquired from an UAS for mapping leaf carotenoid concentration yielding a relative root mean square error (RRMSE) of 14.4%. The estimation of biomass and nitrogen content of wheat and barley were the main objectives in the field study conducted by Pölönen et al. (2013). Spectral feature extraction was performed using three different approaches: spectral indices, spectral unmixing, and spatial features. To estimate biomass and nitrogen content on the field from hyperspectral images, they used machine learning K nearest neighbor classification. The flight campaign was carried out by a lightweight microcopter with autopilot system, which enables it to fly in an advance-determined path. The strongest correlations between predicted and measured values were drawn for biomass prediction when linear model was employed ($R^2 = 0.814$) with neglected advance in comparison to MCARI (Modified Chlorophyll Absorption Ratio Index) indices efficiency ($R^2 = 0.807$), while tissue nitrogen content prediction had 72% confidence by employing linear models.

A low altitude UAS was used by Li et al. (2020) to acquire RGB and hyperspectral imaging data for potato crop at two growth stages to estimate the aboveground biomass and predict crop yield. The images were taken by a lightweight UAS (DJI Phantom 4 Pro) equipped with a 20 MP (megapixel) camera at a flight altitude of 30 m, equivalent to a spatial resolution of 0.5 cm/pixel. The field setup included six cultivars and multiple treatments of nitrogen, potassium, and mixed compound fertilizers. According to the given results, random forest regression models showed high prediction accuracy for fresh and dry aboveground biomass, with a coefficient of determination $(R^2) > 0.90$.

A partial least squares regression model based on the full wavelength spectra demonstrated improved yield prediction $(R^2 = 0.81)$ in comparison to narrowband vegetation indices $(R^2 = 0.63)$. With increased capabilities for the detection of various bands, hyperspectral recording processing becomes more complex. Analysis of hyperspectral imaging data is far more complex. Data at such spectral resolution show multicollinearity that can lead to wider confidence intervals producing less reliable probabilities in terms of the effect of independent variables in a model.

Light detection and ranging

The LIDAR technology is mostly utilized to produce digital elevation models (DEM) or digital terrain models (DTM) for 3D mapping and representation in the 3D world. However, for agriculture purposes, LIDAR has been more employed in phenotyping of crops which is important for characterization of morphological as well as physiological crop traits of varieties of agricultural crops. LIDAR systems have been downsized in recent years, resulting in reduced weights and smaller dimensions, and may thus be operated from UASs. Methods based on multispectral reflectance data with vegetation indices (VI) mostly measure the green area index (GAI) or reaction to chlorophyll content of the canopy surface rather than the full aboveground biomass that may be present from nongreen plant components. The main benefit of LIDAR technologies is that they are sensitive to both green and nongreen plant components (Bates et al., 2021).

LIDAR may be used to estimate morphological characteristics such as plant height, stem diameter, LAI, plant canopy structure, stalk length, and plants' locations. The LIDAR sensors have a distinct advantage over other optical sensors as they are not affected by dust, illumination conditions, or fog. Especially in perennial crops, LIDAR UAS detection has been frequently used in plant height and aboveground biomass estimation (Maesano et al., 2020). Substantial correlation between sensor data and manually measured crop height and biomass were achieved, with determination coefficients of 0.73 and 0.71 for height and biomass, respectively. Bates et al. (2021) performed complex study using LIDAR and multispectral sensors (both mounted on a DJI Matrice 600) in estimating plant area index (PAI) in relation to canopy density throughout a growing season. LIDAR flights were conducted at a height of 50 m aboveground level and at a ground speed of 5 m per second achieving point density 85 pts/sqm with an accuracy of 5 cm. Very promising results in prediction of crop height using LIDAR are demonstrated in studies of Jimenez-Berni et al. (2018) and Sun et al. (2017), who achieved 99% confidence on wheat height detection and 98% confidence on cotton height detection.

ten Harkel et al. (2019) used RIEGL RiCOPTER UAV system (Riegl, Austria) equipped with the VUX-SYS laser scanner to estimate biomass and crop height of winter wheat, sugar beet, and potato. The LIDAR unit has precision of 0.5 cm. To establish a consistent point spacing, the scanner pulse repetition rate was set at 550 kHz, the scanner angle ranged from 30 to 330 degrees, and the scanner speeds were synchronized with the UAS forward speed. UAS-LIDAR data were collected while flying at 40 m aboveground level at a speed of 6 m per second. The accuracy of the plant height measurement using LIDAR varied among 3.4–12 cm depending on the type of crop.

When measuring with LIDAR sensors, the accuracy decreases for crops with a more intricate canopy structure, such as potato and sugar beet, where it is more difficult to offer a high-precision plant height estimate. According to their statements, biomass structure and uniform height pattern, winter wheat is an easy-to-measure crop in the field. Potatoes and sugar beets, as well as their leaf angle distribution, are more difficult to assess in the field, which is exacerbated even more by the fact that potatoes are grown on ridges. Respecting assessment of biomass of the wheat, high reliability was achieved (82%), although significantly less was noted in case of sugar beet. For better performance, the authors suggested combining LIDAR with hyperspectral data.

Conclusion

This chapter covered on overview on the types of UASs and their onboard sensors, mission-planning aspects, and on state-of-the-art technology used in UASs with their application in the field of precision agriculture. UASs equipped with various sensors, such as RGB, thermal, hyperspectral, multispectral cameras, and LIDAR, have become increasingly popular in the field of agriculture. These sensors allow for the collection of detailed, high-resolution data on crop health, soil moisture, and crop yields, among others. These data can be used to optimize crop yields, improve crop health, and reduce the use of pesticides and fertilizers. Overall, UASs equipped with these sensors are a valuable tool in precision agriculture, allowing for more efficient and effective crop management.

References

Addicott, J.E., 2020. The Precision Farming Revolution: Global Drivers of Local Agricultural Methods. Springer Singapore, Singapore. Available from: https://doi.org/10.1007/978-981-13-9686-1.

Avtar, R., Watanabe, T. (Eds.), 2020. Unmanned Aerial Vehicle: Applications in Agriculture and Environment. Springer International Publishing, Cham. Available from: https://doi.org/10.1007/978-3-030-27157-2.

Baluja, J., Diago, M.P., Balda, P., Zorer, R., Meggio, F., Morales, F., et al., 2012. Assessment of vineyard water status variability by thermal and multispectral imagery using an unmanned aerial vehicle (UAV). Irrig. Sci. 30, 511−522. Available from: https://doi.org/10.1007/s00271-012-0382-9.

Barbosa, B.D.S., Araújo e Silva Ferraz, G., Mendes dos Santos, L., Santana, L.S., Bedin Marin, D., Rossi, G., et al., 2021. Application of RGB images obtained by UAV in coffee farming. Remote Sens. 13, 2397. Available from: https://doi.org/10.3390/rs13122397.

Bates, J.S., Montzka, C., Schmidt, M., Jonard, F., 2021. Estimating canopy density parameters time-series for winter wheat using UAS mounted LiDAR. Remote Sens. 13, 710. Available from: https://doi.org/10.3390/rs13040710.

Boesch, R., 2017. Thermal remote sensing with UAV-based workflows. Int. Arch. Photogramm. Remote Sens. Spat. Inf. Sci. 41−46. Available from: https://doi.org/10.5194/isprs-archives-XLII-2-W6-41-2017.

Boreman, G.D., 2005. Classification of imaging spectrometers for remote sensing applications. Opt. Eng. 44, 013602. Available from: https://doi.org/10.1117/1.1813441.

Crusiol, L.G.T., Nanni, M.R., Furlanetto, R.H., Sibaldelli, R.N.R., Cezar, E., Mertz-Henning, L.M., et al., 2020. UAV-based thermal imaging in the assessment of water status of soybean plants. Int. J. Remote Sens. 41, 3243−3265. Available from: https://doi.org/10.1080/01431161.2019.1673914.

Esposito, M., Crimaldi, M., Cirillo, V., Sarghini, F., Maggio, A., 2021. Drone and sensor technology for sustainable weed management: a review. Chem. Biol. Technol. Agric. 8, 18. Available from: https://doi.org/10.1186/s40538-021-00217-8.

Furukawa, F., Laneng, L.A., Ando, H., Yoshimura, N., Kaneko, M., Morimoto, J., 2021. Comparison of RGB and multispectral unmanned aerial vehicle for monitoring vegetation coverage changes on a landslide area. Drones 5, 97. Available from: https://doi.org/10.3390/drones5030097.

Herzig, P., Borrmann, P., Knauer, U., Klück, H.-C., Kilias, D., Seiffert, U., et al., 2021. Evaluation of RGB and multispectral unmanned aerial vehicle (UAV) imagery for high-throughput phenotyping and yield prediction in barley breeding. Remote Sens. 13, 2670. Available from: https://doi.org/10.3390/rs13142670.

Ibarrola-Rivas, M., Kastner, T., Nonhebel, S., 2016. How much time does a farmer spend to produce my food? An international comparison of the impact of diets and mechanization. Resources 5, 47. Available from: https://doi.org/10.3390/resources5040047.

Ivosevic, B., Han, Y.-G., Cho, Y., Kwon, O., 2015. The use of conservation drones in ecology and wildlife research. J. Ecol. Environ. 38, 113−118. Available from: https://doi.org/10.5141/ecoenv.2015.012.

Ivosevic, B., Han, Y.-G., Kwon, O., 2017. Calculating coniferous tree coverage using unmanned aerial vehicle photogrammetry. J. Ecol. Env. 41, 10. Available from: https://doi.org/10.1186/s41610-017-0029-0.

Jimenez-Berni, J.A., Deery, D.M., Rozas-Larraondo, P., Condon, A.(Tony) G., Rebetzke, G.J., et al., 2018. High throughput determination of plant height, ground cover, and above-ground biomass in wheat with LiDAR. Front. Plant. Sci. 9, 237. Available from: https://doi.org/10.3389/fpls.2018.00237.

Kim, J., Kim, S., Ju, C., Son, H.I., 2019. Unmanned aerial vehicles in agriculture: a review of perspective of platform, control, and applications. IEEE Access. 7, 105100−105115. Available from: https://doi.org/10.1109/ACCESS.2019.2932119.

Kostić, M.M., Tagarakis, A.C., Ljubičić, N., Blagojević, D., Radulović, M., Ivošević, B., et al., 2021. The effect of N fertilizer application timing on wheat yield on chernozem soil. Agronomy 11, 1413. Available from: https://doi.org/10.3390/agronomy11071413.

Lacerda, L.N., Snider, J.L., Cohen, Y., Liakos, V., Gobbo, S., Vellidis, G., 2022. Using UAV-based thermal imagery to detect crop water status variability in cotton. Smart Agric. Technol. 2, 100029. Available from: https://doi.org/10.1016/j.atech.2021.100029.

Li, B., Xu, X., Zhang, L., Han, J., Bian, C., Li, G., et al., 2020. Above-ground biomass estimation and yield prediction in potato by using UAV-based RGB and hyperspectral imaging. ISPRS J. Photogramm. Remote Sens. 162, 161−172. Available from: https://doi.org/10.1016/j.isprsjprs.2020.02.013.

Librán-Embid, F., Klaus, F., Tscharntke, T., Grass, I., 2020. Unmanned aerial vehicles for biodiversity-friendly agricultural landscapes - a systematic review. Sci. Total Environ. 732, 139204. Available from: https://doi.org/10.1016/j.scitotenv.2020.139204.

Maesano, M., Khoury, S., Nakhle, F., Firrincieli, A., Gay, A., Tauro, F., et al., 2020. UAV-based LiDAR for high-throughput determination of plant height and above-ground biomass of the bioenergy grass arundo donax. Remote Sens. 12, 3464. Available from: https://doi.org/10.3390/rs12203464.

Mahlein, A.K., Kuska, M.T., Thomas, S., Bohnenkamp, D., Alisaac, E., Behmann, J., et al., 2017. Plant disease detection by hyperspectral imaging: from the lab to the field. Adv. Anim. Biosci. 8, 238−243. Available from: https://doi.org/10.1017/S2040470017001248.

McBratney, A., Whelan, B., Ancev, T., Mcbratney, A., Bouma, J., 2005. Future directions of precision agriculture. Precis. Agric. 17−23. Available from: https://doi.org/10.1007/s11119-005-0681-8.

Messina, G., Modica, G., 2020. Applications of UAV thermal imagery in precision agriculture: state of the art and future research outlook. Remote Sens. 12, 1491. Available from: https://doi.org/10.3390/rs12091491.

Milics, G., 2019. Application of UAVs in precision agriculture. In: Palocz-Andresen, M., Szalay, D., Gosztom, A., Sípos, L., Taligás, T. (Eds.), International Climate Protection. Springer International Publishing, Cham, pp. 93−97. Available from: https://doi.org/10.1007/978-3-030-03816-8_13.

Mogili, U.R., Deepak, B.B.V.L., 2018. Review on application of drone systems in precision agriculture. Procedia Comput. Sci. 133, 502−509. Available from: https://doi.org/10.1016/j.procs.2018.07.063.

Oliver, M.A. (Ed.), 2010. Geostatistical Applications for Precision Agriculture. Springer Netherlands, Dordrecht. Available from: https://doi.org/10.1007/978-90-481-9133-8.

Pádua, L., Vanko, J., Hruška, J., Adão, T., Sousa, J.J., Peres, E., et al., 2017. UAS, sensors, and data processing in agroforestry: a review towards practical applications. Int. J. Remote Sens. 38, 2349−2391. Available from: https://doi.org/10.1080/01431161.2017.1297548.

Pölönen, I., Saari, H., Kaivosoja, J., Honkavaara, E., Pesonen, L., 2013. Hyperspectral imaging based biomass and nitrogen content estimations from light-weight UAV. In: Neale, C.M.U., Maltese, A. (Eds.), Presented at the SPIE Remote Sensing, Dresden, Germany, 88870J. Available from: https://doi.org/10.1117/12.2028624.

Ramesh, P.S., Muruga, L.J., 2020. Mini unmanned aerial systems (UAV) - a review of the parameters for classification of a mini UAV. Int. J. Aviat. Aeronaut. Aerosp. Available from: https://doi.org/10.15394/ijaaa.2020.1503.

Puri, V., Nayyar, A., Raja, L., 2017. Agriculture drones: a modern breakthrough in precision agriculture. J. Stat. Manag. Syst. 20, 507−518. Available from: https://doi.org/10.1080/09720510.2017.1395171.

Roy Choudhury, M., Das, S., Christopher, J., Apan, A., Chapman, S., Menzies, N.W., et al., 2021. Improving biomass and grain yield prediction of wheat genotypes on sodic soil using integrated high-resolution multispectral, hyperspectral, 3D point cloud, and machine learning techniques. Remote Sens. 13, 3482. Available from: https://doi.org/10.3390/rs13173482.

Sun, S., Li, C., Paterson, A., 2017. In-field high-throughput phenotyping of cotton plant height using LiDAR. Remote Sens. 9, 377. Available from: https://doi.org/10.3390/rs9040377.

ten Harkel, J., Bartholomeus, H., Kooistra, L., 2019. Biomass and crop height estimation of different crops using UAV-based LiDAR. Remote Sens. 12, 17. Available from: https://doi.org/10.3390/rs12010017.

Tsouros, D.C., Bibi, S., Sarigiannidis, P.G., 2019. A review on UAV-based applications for precision agriculture. Information 10, 349. Available from: https://doi.org/10.3390/info10110349.

Urbahs, A., Jonaite, I., 2013. Features of the use of unmanned aerial vehicles for agriculture applications. Aviation 17, 170−175. Available from: https://doi.org/10.3846/16487788.2013.861224.

Veroustraete, F., 2015. The rise of the drones in agriculture. EC Agric. 2, 325−327.

Wan, L., Cen, H., Zhu, J., Zhang, J., Zhu, Y., Sun, D., et al., 2020. Grain yield prediction of rice using multi-temporal UAV-based RGB and multispectral images and model transfer − a case study of small farmlands in the South of China. Agric. For. Meteorol. 291, 108096. Available from: https://doi.org/10.1016/j.agrformet.2020.108096.

Wich, S.A., Koh, L.P., 2018. Conservation Drones: Mapping and Monitoring Biodiversity. Oxford University Press.

Wu, D., Sun, D.-W., 2013. Advanced applications of hyperspectral imaging technology for food quality and safety analysis and assessment: a review—part I: fundamentals. Innov. Food Sci. Emerg. Technol. 19, 1−14. Available from: https://doi.org/10.1016/j.ifset.2013.04.014.

Zarco-Tejada, P.J., Guillén-Climent, M.L., Hernández-Clemente, R., Catalina, A., González, M.R., Martín, P., 2013. Estimating leaf carotenoid content in vineyards using high resolution hyperspectral imagery acquired from an unmanned aerial vehicle (UAV). Agric. For. Meteorol. 171−172, 281−294. Available from: https://doi.org/10.1016/j.agrformet.2012.12.013.

Zhang, C., Kovacs, J.M., 2012. The application of small unmanned aerial systems for precision agriculture: a review. Precis. Agric. 13, 693−712. Available from: https://doi.org/10.1007/s11119-012-9274-5.

Zhao, T., Nakano, A., Iwaski, Y., Umeda, H., 2020. Application of hyperspectral imaging for assessment of tomato leaf water status in plant factories. Appl. Sci. 10, 4665. Available from: https://doi.org/10.3390/app10134665.

Applications in agriculture

Application of unmanned aerial systems to address real-world issues in precision agriculture

Bojana Ivošević[1], Marko Kostić[2], Nataša Ljubičić[1], Željana Grbović[1] and Marko Panić[1]
[1]BioSense Institute, University of Novi Sad, Novi Sad, Serbia
[2]Faculty of Agriculture, University of Novi Sad, Novi Sad, Serbia

Introduction

Offering a birds-eye-view, UASs that carry sensors can spot and monitor potential anomalies on the field and reveal many issues such as soil variations, pest infestations, and irrigation problems. On the other hand, flying at low altitude and low speed, UASs offer new opportunities for measurement of agricultural phenomena, delivering fine spatial resolution data at user-controlled revisit periods. Getting closer to the field from above, we might be able to see not just the fields' broad view, but each individual plant.

Collected information from various optical sensors offer rich datasets, such as a range of wavelengths that represents a distinctive spectral signature of plants. With the increasing advancements in electronics, telecommunication, and computation in the last decade, acceleration of UASs introduction in site-specific field observation has been obvious. This technology can be worthwhile either to gain spatiotemporal information, or to execute in-field applications. Indicative applications are specified in the list below and are further analyzed in the subsections that follow:

- Scouting and sensing for nutrient deficiencies
- End-of-season yield and biomass estimation
- Monitoring plant infestation
- Weed detection
- Plant water supply
- Weed control
- Seeding

- Livestock management
- Management zones delineation
- Plant counting

The utilization of UAS in agriculture

Scouting and sensing for nutrient deficiencies

The potential anomalies of the plants or soil could remain undetected when using traditional visual inspection scouting operations. It is important to have a certain level of domain knowledge to understand the biological, physiological, and morphological processes of crops to keep them healthy and manage them in a sustainable and long-term manner. By utilizing UASs, photogrammetry software, and employing analytics, the captured drone images can be stitched together to create highly detailed orthomosaics. These orthomosaics can then be exported as georeferenced vegetation index maps, providing valuable insights at the appropriate scale. Out of many possible applications of unmanned aerial vehicles (UAVs), very common and yet particularly interesting is the acquisition of a field's overview and monitoring the health status of the seasonal vegetation, also known as scouting. For this task, often a relatively low cost UAS with a true color imaging can provide sufficient information (Milics, 2019).

Crop monitoring can be a tricky task, and any field-related issues can be hidden by its abundant foliage while uneven topography often includes areas of soil instability, making it difficult to diagnose and treat nutrient deficiencies before they affect the plants.

Canopy reflectance is used to identify biophysical and biochemical properties of the canopy. The spectral response of the vegetation is unique, as it reflects the plant's health and nutritional status, and is highly dependent on solar radiation, soil properties, and available nutrients. Vegetation indices (VIs) and soil properties can be calculated using optical sensors that can be mounted on a UAV. They are a numerical depiction of the relationship between various wavelengths of light reflected from the plant surface. Researchers derived a vast number of VIs to simplify the monitoring method. Because of the significant correlation between N and absorption of chlorophyll in the visible and near infrared region, most of the VIs were calculated from bands in the visible and near infrared range (Lu et al., 2019a).

Very promising data analysis techniques that are rapidly evolving are related to machine learning (ML) models that can efficiently predict crop health status. Algorithms can segment plant signals for more accurate and earlier detection of plant health issues using the appropriate datasets. Obtaining data at regular intervals creates consistency in data records at targeted growth stages.

To fully understand crop health status, it is necessary to acquire reliable data representing spectral characteristics of plants and correlate them with biophysical and biochemical characteristics collected in traditional manner or in situ measurements. The use of VIs is an immensely popular method. However, some of the indices tend to saturate after a certain level of biomass growth. Vegetative information from remotely sensed images can yield the calculation of a large range of indices affected by the differences and changes of the canopy spectral characteristics.

Balanced plant nutrition is essential for farmer success in agriculture. To achieve that, farmers need a permanent monitoring system which could be able to detect irregularities in plant supply with nutrients from the soil resources. The methods for determination of the nutrient status of plants can either rely on standard laboratory procedures with chemical extraction of plant tissue compounds or using sensing (proximal or remote) technology. Only the rapid and spatially dense observations, such as UAS spectroscopy, fit within the postulates of site-specific field management. Many parameters must be considered such as the application method, the soil type, plants' temporal nutrient requirements, and fertilizer dynamics in the soil. The latter is particularly important as some soil nutrients, like potassium and phosphorus, are firmly bonded to soil particles, while the dynamics of nitrogen in the soil differs (Kostić et al., 2021).

End-of-season yield and biomass estimation

Crop yield projections that are accurate and timely prior to harvest are crucial to the farmers for taking appropriate management decisions. Crop yield is a key determinant of farmers' income and governments' intentions to accomplish food security goals. Agricultural yield forecasts can aid farmers and other stakeholders in better crop planning, such as selling, storing, achieving better market prices, among others. Plant phenotyping and site-specific crop management are made easier with rapid and accurate biomass and yield estimation. Because above-ground biomass is directly connected to crop nutrition and yield, it may be utilized as a crop growth indicator. According to Tagarakis and Ketterings (2017), for better fertilizer in-field management, it is necessary to obtain precise information about site-specific yield potential that would facilitate optimal variable rate application (VRT). Above-ground biomass (AGB) is often used to signify crop growth stages and its status, monitoring and evaluating the applications of agricultural management practices, and the ability to adopt and isolate carbon stocks in agroecosystems (Han et al., 2019; Li et al., 2016). Besides this, it also has an immense contribution in crop breeding and crop yield estimation. Traditional methods for AGB estimation usually use invasive methods including manually sampling, harvesting, and weighing that requires time-consuming and labor-intensive work (Han et al., 2019).

The formulation of an equation to estimate end-of-season yield from mid-season spectral canopy readings is the first step in developing an algo-rithm for variable rate N applications employing spectral sensing. Remote sensing is an effective approach for assessing crop vigor during the growth season and providing information on crop biomass and yield spatial vari-ability. Some researchers derived plant height from a digital surface model from UAS RGB imagery (Li et al., 2020).

The fast and precise estimation of AGB in a noninvasive way is useful for making completely informed crop management decisions. State-of-the-art published papers usually propose a combination of vegetation indices and ML algorithms to obtain precise AGB estimation. In Han et al. (2019), four ML regression algorithms are used and compared: multiple linear regression, support vector machine, artificial neural network, and random forest. Another approach relies on hyperspectral data (Yue et al., 2017). In Zheng et al. (2019), it is shown that the usage of only vegetation indices for estimation of AGB have shown poor performance at high biomass levels. This approach should be improved and expanded by textural and spectral analysis. Image parameters that are used as features are VIs, texture parameters, normalization of texture measurements (normalized difference texture index, NDTI), and combinations of VIs and NDTIs.

Simple linear regression models are often used for final AGB estima-tion. The contribution of VIs has been described in many studies. There are two reasons why dealing with estimates of AGB at different growth stages is a challenging task: (1) vegetation VIs are saturated if the canopy is dense; (2) parts of the plants that grow vertically or are difficult to detect by spectral VIs (Yue et al., 2019). Besides the investigated VIs, canopy height is additionally used to estimate the AGB of various crops. In Lu et al. (2019b), the authors used a low-cost UAS for the evaluation of the utilization of VIs, canopy height, and their combination for AGB estima-tion in wheat. They also used stepwise multiple linear regression (SMLR) and three types of ML algorithms: support vector regression, extreme learning machine, and random forest.

An algorithm developed for coverage assessment, that can be referred to as the estimation of AGB, is a simple and cost-effective solution, using RGB data from UAS. The approach consists of two modules: detecting vegetation, detecting rows and estimating coverage per row. The final evaluation is given as the percentage of vegetation per defined area. Segmentation of vegetation has been done based on VIs and applying appropriate thresholds, often followed by K-means clustering algorithms depending on the type of the crop. For row detection, it is necessary to apply image processing algorithms and mathematical transformations such as Hough Line Transform.

If the goal is to observe the coverage of vegetation per row level, the coverage can be calculated by the percent of the pixels that belong to certain vegetation in the observed region of interest. All these results and conclusions in summary give a valuable outcome that the combination of ML with UAS remote sensing is a promising tool for noninvasive AGB estimation. ML algorithms are well suited for establishing an accurate estimate of wheat biomass. AGB is advantageous for mapping applications and will revolutionize the way we process data and make decisions in agricultural management.

Monitoring plant infestation

Diseases significantly jeopardize crop yield if proper management actions are omitted or delayed. The most important aspect for successful disease control is the timely detection of the infected areas at as early stages as possible. Therefore a well-established plant monitoring system would be particularly valuable. In traditional agriculture, experienced stakeholders, farmers, farm managers or agronomists perform visual inspections, manually scouting for plant diseases. However, modern farms tend to grow larger. In large fields, this method is unsatisfactory due to unreliable human observations which is characterized by time inefficiency and potential misdetection. UAS technology has immense potential for in-field pattern recognition, thanks to the high spatial resolution of the obtained recordings and advanced data mining methods.

By employing UAS, huge areas can be observed rapidly, so, timely decisions can be made which is of crucial importance for pest control. A comprehensive review of Neupane and Baysal-Gurel (2021) is presented for the automatic identification of diseases in different crops using

Figure 3.1 UAS orthomosaics of maize field generated from RGB (left) and multi-spectral (right) cameras. Normalized difference vegetation index calculated from a multispectral camera shows green areas indicating healthy vegetation and orange and red areas as vegetation that needs attention.

the UAS platform in conjunction with deep learning convolutional neural network (CNN) models. The recognition effectiveness varied from 79% for Ramularia blight in cotton to 99% for septorial leaf blight brown spot, frogeye leaf spot, and downy mildew in soybean. These results are considered as very promising.

Various tools are used to process these types of image datasets. Artificial neural networks, decision trees, K-means, k closest neighbors, support vector machines, and regression analysis are some of these techniques. The discrimination among plants may be performed using a variety of vegetative indicators such as normalized difference vegetation index (NDVI) (Fig. 3.1), optimized soil-adjusted vegetation index, and the crop water stress index. Determining the kind of stress, both biotic and abiotic, remains a challenge.

Weed detection

In organic crop production, weeds represent one of the most significant constraints to an efficient agricultural production system (Rajković et al., 2020). Also, conventional crop production is under huge pressure to reduce pesticide consumption in the next decade. The European Commission announced a proposal to restructure the European Union's agriculture sector, making it more sustainable and health-safe (Keating, 2020) which means reducing the use of pesticides by 50% in the next decade.

Additional challenges in crop production arise from actualization of the problem of weed resistance to pesticides. According to Vrbničanin et al. (2017), herbicide resistance has been reported in 478 weed biotypes (252 weed species) in 67 countries. Weed occurrence can have different consequences on the field productivity. In some cases, it jeopardizes product quality (e.g., forage) while in seed production either grain quality or grain yield are endangered. In the context of site-specific weed management, UASs enable quick and precise monitoring of weed presence.

Georeferenced drone recordings from multi/hyperspectral cameras offer wide scale possibilities for weed discrimination and geo-addressing for targeted herbicide application. When it comes to weed detection, the RGB sensor is the most popular payload for UASs. Based on a review study of Esposito et al. (2021) presenting the achievements in UAS for weed detection, some dicotyledonous species (*Amaranthus palmeri*, *Chenopodium album*, and *Cirsium arvense*), as well as several monocotyledonous (*Phalaris* spp., *Avena* spp., and *Lolium* spp.) can be successfully recognized. For wider application by farmers, this technology requires further improvements in terms of robustness of ML algorithms and automatic image processing pipeline.

Plant water supply

Water represents the lifeblood of agriculture ecosystems since it plays important role in many soil properties, biochemical processes, and physiological processes in plants. One of the most important concerns of the 21st century is water shortage due to climate change and inappropriate use of water resources and, in the case of agricultural systems, agrochemicals causing water pollution, having significant impacts on agriculture and increasing water demand (FAO, 2016). Agriculture currently accounts for 70% of worldwide freshwater withdrawals (on average) that imposes needs for urgent optimization of water use.

Due to the yield constraints imposed by water stress, several irrigation scheduling approaches have been developed with the goal of optimizing the time and volume of irrigation water supplied, while also boosting irrigation water usage efficiency (Yu et al., 2017). Leaf water potential (LWP) is widely used for the detection of plant water status as an accurate indicator, but it provides an in situ measurement that limits the number of observations. Similarly, soil moisture can be measured in situ using neutron probe, temporal domain reflectometry (TDR), gravimetric sensors, Aquaterr probe, tensiometers, electrical resistance blocks, or the hand feel technique (Ministry of Agriculture, 2015).

UAS aerial images are successfully used for prediction of soil moisture content (Lu et al., 2020) as well as plant water status (Lacerda et al., 2022).

Sensing the reflectance in the visible light—near infrared, and the thermal infrared spectrum, and using fusion methods combining these two, as well as microwave remote sensing, are some of the remote sensing approaches used to estimate soil moisture.

Weed control

Pesticides are chemicals that are mostly used in agriculture to protect plants from pests, weeds, and diseases. Manually driven and human-carried sprayers, as well as tractor-mounted sprayers, are the primary spraying equipment used in conventional farming. Old fashion pesticide application has been linked to health and environmental problems. From one side, it refers to operator exposure by chemicals and another is pesticide residues that can be detected in a wide range of common foods and drinks, including prepared meals, water, wine, fruit juices, snacks, and animal feeds (Nicolopoulou-Stamati et al., 2016).

The reduction in the use of agrochemicals can be achieved by applying them only when and where they are necessary. Proper management of the spatiotemporal variability of the soil and crop factors of a given field must be taken into consideration. As Chen et al. (2021) mentioned, 80% of UASs will be used for agricultural reasons in the near future, according to the Association for Unmanned Vehicle Systems International (AUVSI). Advantage of low-volume spraying by UAS is supported by the forced air swirling which generates the necessary lift for the airframe, as well as downward airflow that aids spray droplet dispersal into the crop canopy. Unmanned aerial spraying allows for highly accurate and site-specific distribution with less energy and human labor consumption that overlaps with precision agriculture postulates.

Unfortunately, there are several concerns with UAS-based spraying technologies, such as droplet drift and pesticide effectiveness, which must be addressed. Therefore there are many open questions such as: which UAS design (single/multirotor) is most appropriate respecting drift and quality of deposition; battery or petrol-powered drones are most appropriate; what is the most effective flight height; what type of nozzles should be used; what is the best position for the nozzles; what is the most appropriate crop growth stage for each application; and what are the limitations concerning the environmental conditions. Also, many countries do not have clearly defined legislation about UASs application. In most circumstances, one must sift through a plethora of current regulations to understand the regulatory and legal framework in a given nation, industry, and application. According to Chen et al. (2021), for effective UAS spraying, we need more experienced operators who are well versed in the underlying issues that impact UAV spraying quality, not just fundamental UAS operation.

Seeding

Seeding is the most important agrotechnical operation requiring high attention from the executor due to significant impact on uniform plant establishment and further development over vegetational stages. If plants are allowed equal access to all field resources, favorable preconditions for maximizing the genetic potential of the variety-hybrid can be generated (Kostić et al., 2018).

Many other in-field activities are in relation to the seed spatial and vertical distribution. Depending on the crop, the relevance of plant distribution in the field in terms of quality and yield size varies. The multirow seeders are commonly used in conventional cropping systems which impose tractor engagement and power seeder devices. This seeding configuration has shortcomings in terms of limited effects on different terrain conditions and soil conditions deterioration affected by wheel compression.

Aerial seeding would be a useful alternative to ground planting, giving a solution for field crops, such as rapeseed, to be planted in areas where ground machines are inaccessible or simply have poor economic efficiency (Huang et al., 2020). Aerial planting capabilities have been added to the agricultural UASs of several Chinese tech companies to make them more flexible and acceptable. All proposed seeding solutions related to UAS platforms are equipped with an air-assisted centralized seed metering device for wide-spreading in chaotic spatial order. Precision seeding using UASs has not been observed yet due to very tough criteria which must be satisfied such as high accuracy of in-row distance of single-seeded seed and uniform seeding depth.

Livestock management

Because free range animals, such as cattle and sheep, spend a lot of time in the open farm environments, fields, pastures, hills, and moorland, in most cases, keeping track of their location is time consuming and labor intensive. Traditional tracking techniques rely on human observations to identify individual animals, either by natural qualities like coloration and breed, or by artificial measures such as tags or painted-on symbols.

This might lead to observer weariness and mistakes. Drones can successfully monitor and move animals on the farm, regardless of the size of the herd or the fields on which they graze. The next natural step in agricultural technology is the use of UASs for livestock management. Drones with cameras and infrared imaging capabilities can track livestock movement and health, count cattle, and check their behavior. Unmanned aircrafts can find stray cattle and sheep, as well as detect ill or injured animals, reducing the need for people on the ground and improving the information accessible to farmers for livestock management. Weight, size, and physical activity may all be measured with UASs. According to Yaxley et al. (2021), integrating auditory signals with drones can increase safety without putting the animals under stress.

Management zone delineation

A management zone is a productive unit of the field with relatively high internal homogeneity that separates it from other field areas in terms of unique soil qualities that affect productivity. The spatial information, that is usually mentioned as the most helpful, should be merged with the yield through quantitative or densely relation, and in that way, the zone management delineation can be defined as a way of spatially classifying the variability within the field. The objective to delineate management zones within the fields has a various usage in devising strategies including precision and variable fertilization, irrigation, and harvest (Starý et al., 2020).

Also, its importance lies in maximizing economic return by optimizing rates of yield-limiting inputs or monitoring the damaging effects of weed occurrence and other crop pests. The high resolution of images provided by UASs allows agricultural managers and workers to quickly spot and evaluate if any spatial and temporal variability in a field exists. If such a situation does occur, specific areas that would need additional work and monitoring appear, including areas that may have changed in appearance, showing increased, or decreased quality, or having other issues with the crops' condition. VIs reveal important soil and crop properties that affect growth and productivity and facilitate delineation of management zones. This allows agricultural management workers to quickly act accordingly based on UAS data (e.g., additional fertilization and watering). Approaches based on mathematical procedures that perceive spatial patterns across time series of yield maps can be used as modules in novel tools for detecting intervisibility within regions in the field (Doerge, n.d.).

For the delineation of management zones, classification-based methods are intuitively a good approach with the aim to find the optimal number of clusters that minimize intraclass variance. With additional image processing, the objective is to achieve the most uniform clusters, declared to be separate classes, which involve all the management zones that require the same management strategy. These methods do not take into consideration the spatial information, which imposes a constraint to declare the obtained clusters as zones (CoxSunsetBeach, 2019). Classic ML algorithms, such as K-means or fuzzy c-means are the most common techniques for deriving management zones (Castrignanò et al., 2015; Ouazaa et al., 2020; Santos et al., 2019; Silva et al., 2017).

UAS-based generated orthomosaics can serve as appropriate input data to the clustering algorithm for delineation of management zones (Fig. 3.2). The workflow for the delineation of management zones based on UAV-derived images is as follows: after the separation of the image bands, certain VIs are calculated, and threshold is specified and chosen as a criterion for zones delineation. After these preprocessing steps, K-means results labels that are proclaimed as zones. Considering that homogeneity within zones is one of the main objectives, additional postprocessing steps are performed, in order to achieve as much as possible uniformity within the resulted zones.

Figure 3.2 Management zones delineation at a blueberries field, in Babe, Serbia; three management zones were obtained interpreted by different colors.

The obtained results have various uses in further analysis of the most valuable factors that affect yield or growth. RGB cameras can be a good alternative to multispectral mainly due to their lower price and the possibility of calculating many VIs using appropriate equations from RGB images. VIs obtained from the three basic channels have various usage. Indices such as excess green (ExG), excess green−red (ExGR), or color index of vegetation extraction (CIVE) obtained from RGB images were used to distinguish vegetation from the soil which is important for classification and extraction of the plants (Kataoka et al., 2003).

Rouze et al. (2021) used several indices, derived using RGB images, as a criterion for management zone delineation: ExG, CIVE, triangular greenness index (TGI), and green leaf index (GLI). The equations for the calculation of each index are provided below.

ExG:

$$2 \times \rho G - \rho R - \rho B$$

CIVE:

$$0.441 \times \rho R - 0.881 \times \rho G + 0.385 \times \rho B + 18.787$$

TGI:

$$0.5 \times [(\lambda R - \lambda B)(\rho R - \rho G) - (\lambda R - \lambda G)(\rho R - \rho B)]$$

GLI:

$$(2 \times \rho G - \rho R - \rho B)/(2 \times \rho G + \rho R + \rho B)$$

where ρB, ρG, and ρR are the reflectances at the blue, green, and red bands. Additionally, for TGI, λB, λG, and λR are the wavelengths of the blue, green, and red bands (Saponaro et al., 2021).

Numerous other indices are utilized with the scope to capture field, soil, and crop properties and serve as parameters for the delineation of management zones. However, these require the utilization of multispectral or hyperspectral sensors, which require increased investment. Apart from the sensing tasks, for developing usable and practical zone maps, there are additional requirements that need to be considered, such as the spatial footprint of the machine, or time lags of machinery.

Plant counting

Identifying the optimal in-row spacing and monitoring how many plants successfully emerged is a critical management aspect for crop production.

Plant counting can be achieved using image recognition methods. Estimating the plant number per hectare is an important index for assessing plant density and is directly linked with yield estimation. With the aid of artificial intelligence (AI) and image morphology algorithms, we are able to determine ground coverage and detect plant numbers in different spacing. Using UAS images to detect ground coverage and determine plant number is a challenging task since it requires high image resolution, which in some cases can only be achieved with low altitude flights.

Accompanying unpredictable climatic changes, farmers are forced to pay more attention and make an additional effort for each seed to reach its full potential. Conventional methods and analysis are usually labor intensive, time consuming, and prone to human error; these include manual in-field actions such as: counting the number of plants within a certain range, then estimate the average plant population over the field, counting plants within a certain predefined surface that after merging with appropriate charts refers to the average plant number based on the surface are, or using ruler on the chosen length, counting plants into two rows, then calculating average emergence count (Automated Plant Counts with Drones for Precision Agriculture, 2019).

Even though these traditional methods have a long history, the estimation of plant number is not complete without a clear insight into how many plants have not emerged and where those seeds are. Involving UASs in the framework of precision agriculture for crop monitoring, plant counting, especially at germination and the initial phenological stages of the plants, is rapid and easy to perform and can be much more accurate than the traditional estimations. The result of plant counting can be used in many further analyses and applications within farmers' timely actions in preventing yield losses, such as: plant emergence and crop population evaluation, seed quality estimation. Also, management actions can be decided such as replanting in regions with poor emergence that would possibly show low yield, and monitoring the performance of the replanting actions.

Apart from plant counting, plant density ratio is also an important parameter because it provides fast, valuable information, saves time and resources (cost efficient), provides real-time data that can be used in timely decision making, allowing for corrective measures. According to the state of the art, precision of this monitoring actions is high, showing small error (less than 2%) representing accuracy at the level of centimeter.

A case of using segmentation algorithm for plant counting was setup in a maize experiment in Ravno Selo in the Republic of Serbia. The analysis relies on the image processing detecting the morphology within images and the method consists of two steps/modules: segmentation of vegetation and plants detection and counting (Fig. 3.3).

Segmentation, as one of the most challenging tasks in image processing, has an immense impact on the accuracy of this algorithm. Vegetation recognition can be impacted by occurrence of weeds. Therefore the UAS operators as well as data and image analysts need to have a clear insight of the field, to decide which technique to use depending on the status of the crop and the field. If the plants are at early growth stage, and visibly separated from each other, the detection and counting comes down to detection of centroid of the plant, observing the plants as components of the image. In the cases where plants are overlapped, the additional separation needs to be conducted, with the aim to achieve the highest accuracy of counting.

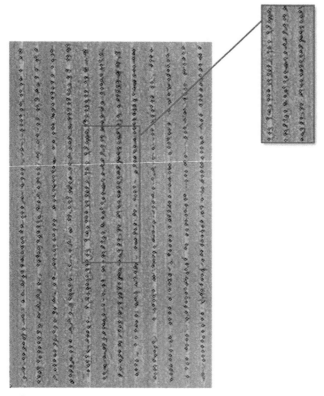

Figure 3.3 Plant counting segmentation on UAS-acquired image from a maize experiment setup in Ravno Selo, Serbia.

The criteria for vegetation segmentation are usually based on setting the appropriate thresholds of vegetation indices. Furthermore, K-means or similar ML algorithms are utilized for clustering. After detecting vegetation, independent components, such as the area, height, and width of the plants are analyzed. The plants' centroids are defined and the distance among them is calculated, for precise counting.

If plants are not easy to segment, even at early growth stages, the agronomist specialist needs to assist for that purpose. Some relevant state-of-the-art published work has been based on UAS-derived data and image processing using ML algorithms. Valente et al. (2020) proposed a method relying on VIs, Otsu threshold, and CNN. In Valente et al. (2020) and Oh et al. (2020), plant counting was obtained using deep learning techniques, but both had limitations due to the process of labeling images in a given dataset to train ML models. The main goal of using UASs and data analytics via ML algorithms is to represent valuable information without requiring technical expert knowledge. The implementation of such algorithms in one information system can have numerous benefits; fast and comprehensive insight of the entire field and automated quantification of crop properties such as plant density, distance, canopy size, or occurrence of gaps that might require immediate attention.

Future perspective of UAS in precision agriculture

The use of remote sensing techniques in precision agriculture is still limited. In addition, there is a lack of information to farmers regarding UAS benefits. Thus UAS is still in its infancy stage in the framework of applying precision agriculture, but there is a substantial space for future development. A key to the success of precision agriculture is to encourage farmers to consider these devices and systems as future tools in their efforts to increase yields and profits. The possibilities of UASs are endless, especially when combined with in situ measurements, image-processing techniques, and deep-learning algorithms.

It is a common practice to use numerous drones instead of one platform for various tasks. In the future, multitasking of UAV platforms can be applied in precision agriculture for performing various missions. The concept of using drone swarms can perform the task faster and more efficiently. This technology uses AI and is still in the development phase. The application of AI will enable the drones to make smart decisions and act independently without the need for human supervision.

The use of drones in agriculture is still not practiced as extended, lies in the insufficient knowledge for the majority of farmers concerning this technology, the increased cost of equipment and the complicated and time-consuming data processing procedures. Furthermore, the use of drones is still dependent on weather conditions, mostly wind and precipitation. In addition to the above, the use of sensors such as various multispectral/hyperspectral/thermal cameras bring with it certain challenges. The quality of aerial images is subjected to atmospheric influence, intensity of solar radiation, position of the sun during the day, and wind speed.

One of the most important aspects affecting the application of drones is the legal provisions that regulate their use. In many countries, regulatory standards are not best suited to the needs of drones. Countries such as the United States, Germany, Great Britain, Spain, and the Netherlands have the clearest and best regulated system of laws and regulations on UASs, while other countries lag in that regard. The cost of the relevant equipment is also one of the major barriers to wider acceptance of such systems in agriculture. Although the arguments relate to the long-term positive effects of the application, it is very difficult to convince potential users to invest in this technology.

According to Sylvester (2018), farmers themselves and their potential to face the upcoming technological changes are most responsible for the modernization of agriculture. In the technological developments of UASs and other sensing systems dedicated to agricultural use, there is a trend for improving user-friendliness of such systems. Spectral UAS sensing, in this regard, necessitates a thorough understanding of all aspects of its use, from survey preparation to production of final orthomosaics and decision-making information. Fortunately, with the advancements of UAS hardware and software, the improved data processing techniques, and updated UAS aviation regulations, in combination with the lower costs, a greater adoption of UAS-based remote sensing and precision agriculture is expected.

References

Automated Plant Counts with Drones for Precision Agriculture [WWW Document], 2019. Available from: https://www.precisionhawk.com/blog/media/topic/understanding-your-aerial-data-plant-counting (accessed 1.23.22).

Castrignanò, A., Landrum, C., De Benedetto, D., 2015. Delineation of management zones in precision agriculture by integration of proximal sensing with multivariate geostatistics. Examples of sensor data fusion. Agric. Conspec. Sci. 80, 39—45.

Chen, P., Ma, X., Wang, F., Li, J., 2021. A new method for crop row detection using unmanned aerial vehicle images. Remote Sens. 13, 3526. Available from: https://doi.org/10.3390/rs13173526.

CoxSunsetBeach, 2019. Delineation of management classes and management zones in Precision Agriculture [WWW Document]. Aspexit. Available from: https://www.aspexit.com/delineation-of-management-classes-and-management-zones-in-precision-agriculture/ (accessed 1.23.22).

Doerge, T.A., n.d. Management Zone Concepts (SSMG-2) 4.

Esposito, M., Crimaldi, M., Cirillo, V., Sarghini, F., Maggio, A., 2021. Drone and sensor technology for sustainable weed management: a review. Chem. Biol. Technol. Agric. 8, 18. Available from: https://doi.org/10.1186/s40538-021-00217-8.

FAO, 2016. Coping with water scarcity in agriculture: a global framework for action in a changing climate.

Han, L., Yang, G., Dai, H., Xu, B., Yang, H., Feng, H., et al., 2019. Modeling maize above-ground biomass based on machine learning approaches using UAV remote-sensing data. Plant Methods 15, 10. Available from: https://doi.org/10.1186/s13007-019-0394-z.

Huang, X., Zhang, S., Luo, C., Li, W., Liao, Y., 2020. Design and experimentation of an aerial seeding system for rapeseed based on an air-assisted centralized metering device and a multi-rotor crop protection UAV. Appl. Sci. 10, 8854. Available from: https://doi.org/10.3390/app10248854.

Kataoka, T., Kaneko, T., Okamoto, H., Hata, S., 2003. Crop growth estimation system using machine vision, in: Presented at the Proceedings 2003 IEEE/ASME International Conference on Advanced Intelligent Mechatronics (AIM 2003), pp. b1079-b1083 vol.2. Available from: https://doi.org/10.1109/AIM.2003.1225492.

Keating, D., 2020. Available from: https://www.forbes.com/sites/davekeating/2020/05/20/eu-plans-to-reduce-pesticides-by-50/.

Kostić, M., Rakić, D., Radomirović, D., Savin, L., Dedović, N., Crnojević, V., et al., 2018. Corn seeding process fault cause analysis based on a theoretical and experimental approach. Comput. Electron. Agric. 151, 207−218. Available from: https://doi.org/10.1016/j.compag.2018.06.014.

Kostić, M.M., Tagarakis, A.C., Ljubičić, N., Blagojević, D., Radulović, M., Ivošević, B., et al., 2021. The effect of N fertilizer application timing on wheat yield on chernozem soil. Agronomy 11, 1413. Available from: https://doi.org/10.3390/agronomy11071413.

Lacerda, L.N., Snider, J.L., Cohen, Y., Liakos, V., Gobbo, S., Vellidis, G., 2022. Using UAV-based thermal imagery to detect crop water status variability in cotton. Smart Agric. Technol. 2, 100029. Available from: https://doi.org/10.1016/j.atech.2021.100029.

Li, W., Niu, Z., Chen, H., Li, D., Wu, M., Zhao, W., 2016. Remote estimation of canopy height and aboveground biomass of maize using high-resolution stereo images from a low-cost unmanned aerial vehicle system. Ecol. Indic. 67, 637−648. Available from: https://doi.org/10.1016/j.ecolind.2016.03.036.

Li, B., Xu, X., Zhang, L., Han, J., Bian, C., Li, G., et al., 2020. Above-ground biomass estimation and yield prediction in potato by using UAV-based RGB and hyperspectral imaging. ISPRS J. Photogramm. Remote Sens. 162, 161−172. Available from: https://doi.org/10.1016/j.isprsjprs.2020.02.013.

Lu, F., Sun, Y., Hou, F., 2020. Using UAV visible images to estimate the soil moisture of steppe. Water 12, 2334. Available from: https://doi.org/10.3390/w12092334.

Lu, N., Wang, W., Zhang, Q., Li, D., Yao, X., Tian, Y., et al., 2019a. Estimation of nitrogen nutrition status in winter wheat from unmanned aerial vehicle based multiangular multispectral imagery. Front. Plant Sci. 10, 1601. Available from: https://doi.org/10.3389/fpls.2019.01601.

Lu, N., Zhou, J., Han, Z., Li, D., Cao, Q., Yao, X., et al., 2019b. Improved estimation of aboveground biomass in wheat from RGB imagery and point cloud data acquired with a low-cost unmanned aerial vehicle system. Plant Methods 15, 17. Available from: https://doi.org/10.1186/s13007-019-0402-3.

Milics, G., 2019. Application of UAVs in precision agriculture. In: Palocz-Andresen, M., Szalay, D., Gosztom, A., Sípos, L., Taligás, T. (Eds.), International Climate Protection. Springer International Publishing, Cham, pp. 93−97. Available from: https://doi.org/10.1007/978-3-030-03816-8_13.

Ministry Of Agriculture, 2015. Irrigation Scheduling Techniques.

Neupane, K., Baysal-Gurel, F., 2021. Automatic identification and monitoring of plant diseases using unmanned aerial vehicles: a review. Remote Sens. 13, 3841. Available from: https://doi.org/10.3390/rs13193841.

Nicolopoulou-Stamati, P., Maipas, S., Kotampasi, C., Stamatis, P., Hens, L., 2016. Chemical pesticides and human health: the urgent need for a new concept in agriculture. Front. Public Health 4. Available from: https://doi.org/10.3389/fpubh.2016.00148.

Oh, S., Chang, A., Ashapure, A., Jung, J., Dube, N., Maeda, M., et al., 2020. Plant counting of cotton from UAS imagery using deep learning-based object detection framework. Remote Sens. 12, 2981. Available from: https://doi.org/10.3390/rs12182981.

Ouazaa, S., Barrero, O., Quevedo Amaya, Y.M., Chaali, N., Montenegro Ramos, O., 2020. Site-specific management zones delineation and Yield prediction for rice based cropping system using on-farm data sets in Tolima (Colombia) 2466.

Rajković, M., Malidža, G., Stepanović, S., Kostić, M., Petrović, K., Urošević, M., et al., 2020. Influence of burner position on temperature distribution in soybean flaming. Agronomy 10, 391. Available from: https://doi.org/10.3390/agronomy10030391.

Rouze, G., Neely, H., Morgan, C., Kustas, W., Wiethorn, M., 2021. Evaluating unoccupied aerial systems (UAS) imagery as an alternative tool towards cotton-based management zones. Precision Agric. 22, 1861−1889. Available from: https://doi.org/10.1007/s11119-021-09816-9.

Santos, S.G., Melo, J.C., Constantino, R.G., Brito, A.V., 2019. A Solution for Vegetation Analysis, Separation and Geolocation of Management Zones using Aerial Images by UAVs, in: Presented at the 2019 IX Brazilian Symposium on Computing Systems Engineering (SBESC), pp. 1−8. Available from: https://doi.org/10.1109/SBESC49506.2019.9046079.

Saponaro, M., Agapiou, A., Hadjimitsis, D.G., Tarantino, E., 2021. Influence of spatial resolution for vegetation indices' extraction using visible bands from unmanned aerial vehicles' orthomosaics datasets. Remote Sens. 13, 3238. Available from: https://doi.org/10.3390/rs13163238.

Silva, G., Escarpinati, M., Abdala, D., Souza, I., 2017. Definition of Management Zones Through Image Processing for Precision Agriculture. Available from: https://doi.org/10.1109/WVC.2017.00033.

Starý, K., Jelínek, Z., Kumhálová, J., Chyba, J., Balážová, K., 2020. Comparing RGB - based vegetation indices from UAV imageries to estimate hops canopy area. Available from: https://doi.org/10.15159/ar.20.169.

Sylvester, G., 2018. E-Agriculture in Action: Drones for Agriculture. Food and Agriculture Organization of the United Nations and International Telecommunication Union, Bankok.

Tagarakis, A.C., Ketterings, Q.M., 2017. In-season estimation of corn yield potential using proximal sensing. Agron. J. 109, 1323−1330. Available from: https://doi.org/10.2134/agronj2016.12.0732.

Valente, J., Sari, B., Kooistra, L., Kramer, H., Mücher, S., 2020. Automated crop plant counting from very high-resolution aerial imagery. Precis. Agric. 21, 1366−1384. Available from: https://doi.org/10.1007/s11119-020-09725-3.

Vrbničanin, S., Pavlović, D., Božić, D., 2017. Weed resistance to herbicides. In: Pacanoski, Z. (Ed.), Herbicide Resistance in Weeds and Crops. InTech. Available from: https://doi.org/10.5772/67979.

Yaxley, K.J., Joiner, K.F., Abbass, H., 2021. Drone approach parameters leading to lower stress sheep flocking and movement: sky shepherding. Sci. Rep. 11, 1−9. Available from: https://doi.org/10.1038/s41598-021-87453-y.

Yue, J., Yang, G., Li, C., Li, Z., Wang, Y., Feng, H., et al., 2017. Estimation of winter wheat above-ground biomass using unmanned aerial vehicle-based snapshot hyperspectral sensor and crop height improved models. Remote Sens. 9, 708. Available from: https://doi.org/10.3390/rs9070708.

Yue, J., Yang, G., Tian, Q., Feng, H., Xu, K., Zhou, C., 2019. Estimate of winter-wheat above-ground biomass based on UAV ultrahigh-ground-resolution image textures and vegetation indices. ISPRS J. Photogramm. Remote Sens. 150, 226−244. Available from: https://doi.org/10.1016/j.isprsjprs.2019.02.022.

Yu, K., Kirchgessner, N., Grieder, C., Walter, A., Hund, A., 2017. An image analysis pipeline for automated classification of imaging light conditions and for quantification of wheat canopy cover time series in field phenotyping. Plant Methods 13, 15. Available from: https://doi.org/10.1186/s13007-017-0168-4.

Zheng, H., Cheng, T., Zhou, M., Li, D., Yao, X., Tian, Y., et al., 2019. Improved estimation of rice aboveground biomass combining textural and spectral analysis of UAV imagery. Precis. Agric. 20, 611−629. Available from: https://doi.org/10.1007/s11119-018-9600-7.

CHAPTER 4

Unmanned aerial vehicles applications in vegetables and arable crops

Vasilis Psiroukis, George Papadopoulos, Nikoleta Darra,
Michael Gerasimos Koutsiaras, Ari Lomis, Aikaterini Kasimati and
Spyros Fountas
Department of Natural Resources Management and Agricultural Engineering, Agricultural University of
Athens, Athens, Greece

Introduction

The recent boom in agricultural data caused by the widespread accessibility of sensing technologies in the sector has enabled complex monitoring methodologies to be easily adopted, providing constant data streams generated from field-specific multiple sensing systems throughout the cultivation period. Unoccupied (or unmanned) aerial vehicles (UAVs) are the most recent type of platforms dedicated to remote sensing, which in recent years have become a hot spot for both research and commercial applications across the agricultural sector. The main reason behind the widespread use and interest for UAVs in agriculture is their wide range of applications along with the unparalleled efficiency and flexibility they offer for a variety of situations (Tsouros et al., 2019; Yang et al., 2017). They have a unique capacity to cover large areas, provide data of very high spatial resolution and at a high temporal frequency, without damaging the growing environment or disrupting the observed ecosystems. For these reasons, UAVs have also become heavily utilized tools for many emerging applications in the area of ecological monitoring and biodiversity conservation, domains which are closely related to agriculture. UAV remote sensing provides a nondestructive and cost-effective way for rapid monitoring of agricultural fields using accessible and cost-effective platforms, capable of flying at low altitudes, and capturing images of unparalleled spatial resolution (Matese et al., 2015). Moreover, as the flights are performed by human operators on demand and are much easier to be deployed than manned aircrafts, UAVs offer a potentially very high

Unmanned Aerial Systems in Agriculture
DOI: https://doi.org/10.1016/B978-0-323-91940-1.00004-9

temporal resolution. The only barriers are the weather conditions that may not allow for a safe operation (Burkart et al., 2015).

The choice of acquisition timing and frequency is an essential characteristic for multiple agricultural applications with strict time windows. Moreover, the ability of UAVs to acquire data close to the surface and the integration of correction systems (i.e., sunlight correction sensors) makes UAV imagery collection unaffected by cloud coverage (Berni et al., 2009), similarly to manned aerial flights. Another advantage of UAVs is their capacity to carry multiple sensors, as different sensing systems can easily be mounted and unmounted through aircraft modifications and gimbals. Naturally, different sensors cannot perform optimal data collection during a single mission flight, as their required flight parameters may vary, thus, individual flights for each sensor may be demanded. Nevertheless, the possibility of a single aircraft being capable of generating multiple datasets is still a major advantage, especially when taking into consideration the time efficiency of UAV flights and the relatively low acquisition costs of both the aircrafts and their sensing components. The capabilities of UAVs in agriculture, however, are not limited to just locating problematic areas. They are useful in monitoring several crop growth parameters throughout the season, while the emergence of data science and computer vision applications in agriculture has enabled very sophisticated applications, such as crop stress and disease identification, weed infestation mapping, and plant-level yield and maturity identification.

Sensing and surveying

Plants have a special behavior in terms of light reflectance, as photosynthetically active tissues demonstrate high absorption of red light, while reflecting the largest portion in the green spectrum—and therefore appear green to us (Tucker, 1979). Outside from the spectrum that we can perceive with our eyes, plants also reflect most of the near-infrared (NIR) light that hits their canopy, due to the spongy mesophyll (Tucker, 1979). If this is not the case, it is an indicator that something prevents the plants from performing their biological functions properly. Such factors can vary, from a source of stress, to a disease infestation (Tucker, 1980). These deviations can be detected quicker by identifying variations in reflectance, before they become visible to the human eye, via specific symptoms such as chlorosis, when the leaves no longer appear green to us. This reflectance behavior or "profile" is called spectral signature, and is the basis for remote sensing applications in agriculture as it can provide direct information about crop health and overall condition. Remote sensing leverages on this spectral signature to estimate the properties of plants through non-destructive processes in a fast and accurate way (Moran et al., 1997), and has a wide range of applications in agriculture, including but not being limited to crop growth monitoring, crop yield and quality estimation, identification of irrigation needs, as well as biotic and abiotic damage such as pests infestations, disease infections, hail damage, flood and drought damage (Mondal and Basu, 2009).

The characteristic property of the spectral signature and the potential it held in all these applications in agriculture sparked the rapid development of numerical indices resulting from simple formulas using the percentage reflections in selected "critical" spectra. These indices are called vegetation indices (VIs), and are mathematical quantitative combinations of the absorption and scattering rates of plants in different bands of the electro-magnetic spectrum used to detect, quantify, and monitor specific crop-related parameters. VIs provide a simple yet elegant method for measuring plant responses throughout the season, exploiting the basic differences between soil and plant spectra, and are often calculated as a type of rela-tionship between reflected light in the visible and NIR wavelengths. Many such indices were rapidly created to cover a variety of different applications since satellite imagery became widely accessible, starting in 1972 (Cihlar et al., 1991), with the launch of Landsat 1. In 1973, how-ever, Rousse introduced the normalized difference vegetation index (NDVI), calculated as the ratio of the difference and the sum of the reflectance in the NIR and red bands, which remains the most widely used index to this day. As of today, several hundreds of VIs have been developed to address specific problems in the agricultural sector alone, such as crop monitoring, disease detection, irrigation and input applica-tions optimization, crop identification, as well as yield estimation and maturity identification.

In agricultural applications, UAVs can carry a varied array of sensors and electronics, ranging from image acquisition sensors (visible, multispec-tral, and hyperspectral cameras) to active sensing systems (e.g., light detec-tion and ranging—LiDAR) and integrated automation mechanisms (e.g., real-time actuators or spraying systems). They are equipped with on-board positioning system devices and stabilization sensors (mainly three-axis gyroscopes), allowing them to fly autonomously given a certain set of predefined commands from the ground control station (GCS). This set of commands is usually a sequence of points in the real world (waypoints) that the aircraft is tasked to navigate toward using its internal positioning and orientation systems. Each waypoint, which is perceived by the UAV as a set of orders, includes not only the coordinates of the waypoint, but also other flight parameters, such as cruising speed and altitude. These parameters can be either fixed for the entire duration of the mission flight, or change from point to point. Moreover, certain actions are also provided in the form of orders, such as take-off and landing, as well as the signal to trigger certain components (e.g., the sensing device to capture an image) whenever there is an established link with the UAV. The UAV 2D mapping process involves the collection of individual images to create orthophotogrammetric (or orthomosaic) maps. The photogrammetric process essentially "stitches" each individual image with its neighbors, based on georeference metadata and their extracted common features (the common points between neighboring images).

To achieve this, the generated dataset should guarantee that the sequence of aerial images meets the photogrammetric requirements and there are enough corresponding feature points to perform image mosaicking. To this end, a number of flight parameters related to the function of both the aircraft and the sensing device should be determined and combined in a delicate way to ensure that the dataset is eligible for mosaicking. The first parameter that is often considered when designing a UAV surveying mission is the sensing instrument's field of view. It is expressed in two angle values, representing its vertical and horizontal field of view, respectively. Through these values, we can easily calculate the area covered by the sensing instrument in each image capture, by calculating the triangular values of these angles and the perpendicular side of the aircraft's altitude.

Image overlap is the parameter that ensures that sufficient common features between the dataset's images exist, and therefore high-quality mosaicking is achievable (Haala and Rothermel, 2012). In aerial imagery, there are two overlap parameters: front overlap (or frontlap) and side overlap (or sidelap). In agricultural surveys, UAVs normally move in straight lines, called flight-lines. Frontal overlap represents the overlapping percentage of consecutive images captured on the same flight-line. Sidelap refers to the percentage of overlap between different flight-lines, or simply the common area scanned between neighboring images of consecutive flight-lines. Overlaps are the only parameter that are affected by all other flight parameters, both related to the aircraft (i.e., speed and altitude) and the sensing device (i.e., camera field of view and capture interval) (Xing et al., 2010). As a result, they are considered a standard requirement, and are the first parameter to be taken into consideration. A common practice for nadir flights (the most common type of flights in agricultural surveys, where the flight-lines are parallel to the earth's surface and data are collected vertically, with the camera's field of view being perpendicular to the ground), a standard practice is to design flight plans with 70% and 80% frontal and side overlaps, respectively (Ni et al., 2018), as higher values might result in unnecessary large volumes of data and higher processing times at the next stages of data processing.

Ground sampling distance (GSD) is a critical parameter in all drone mapping and surveying projects. GSD is defined as the distance between the centers of two adjacent pixels measured in the real world (the ground, in nadir flights). Essentially, GSD translates distances in the final orthomosaic to actual distances on the ground. It is described in side length per pixel (cm/px), and is related to the camera's focal length and resolution (number and arrangement of pixels), and of course the camera's distance from the surveyed area (altitude). The further the camera is from the target, the less "space" the target will occupy on the image. GSD scales linearly with the drone's altitude if it is carrying a camera with a fixed focal length lens, and linearly with focal length if the camera has a variable focal length lens. Naturally, lower altitude flights improve the GSD and smaller objects become more distinguishable in the final orthomosaic. GSD is highly dependent on each specific survey and application. For simple crop monitoring tasks, a higher GSD is acceptable (i.e., 5–10 cm/px). For tasks where smaller objects should be clearly visible to the final image, however, GSD should be considerably higher (i.e., lower than 2 cm/px), and the entire flight plan should be designed to ensure this value.

Altitude and speed are the two aircraft-related flight parameters that must be properly balanced to produce high-resolution imagery with sufficient overlap values. Lower altitude flights generate data of higher resolution (GSD), however, the aircraft must travel at a lower speed to avoid distortion and motion blur, as well as to maintain sufficient frontal overlap. Moreover, the flight-lines should become denser, to ensure sufficient side overlap, as the camera moves closer to the ground and the area scanned with its default field of view is reduced. As a result, flight duration increases significantly, and often the operators are met with challenging situations where the data quality and flight efficiency trade-off should be optimized (Psiroukis et al., 2021).

The final parameter of flight planning is the data capturing method of the UAV system. Two different approaches exist based on whether the UAV has an established link with the data capturing device (camera sensor). The first method is the most common one, when there is no established link between the sensing device and the aircraft. In this case, the sensing instrument is set to capture data at a fixed interval, and the entire flight mission is designed around this parameter. This case perfectly represents one of the very first definitions given to aerial photography, referring to it as "a means of fixing time within the framework of space." The second method involves a set communication between the aircraft and the camera. In this case, the entire flight plan is designed to achieve the desired parameters (overlaps and GSD), and the capturing interval is calculated last, as the system can signal the sensor on when to capture each image, essentially adjusting the capture speed in real time. The signal is either triggered when the UAV has traveled a certain distance since the start of each flight-line or the previous capture, or a certain time between captures is calculated at the start of each flight-line (according to the predefined overlaps) and the sensor captures in this interval.

Finally, another parameter that should be considered when designing a UAV flight mission is aircraft stability. This parameter cannot be quantified by a certain metric, but should be always optimized in an attempt to maximize the quality of the collected data. Drones have known operational limits, set by the manufacturer, corresponding to the environmental conditions under which the aircraft can operate safely. High wind speeds and the presence of gusts and turbulence are the most common factors for flight cancellations. Even within the safe operational limits, however, the UAV might be able to successfully complete the mission, but the quality of data might not be as high as expected. This is because aircraft stability is a crucial factor for collecting high-quality images, and it is heavily impacted by wind parameters. Collision with gusts can tilt the aircraft and the sensing system in certain angles, deviating from its nadir position and resulting in data captured from an oblique angle. This causes distortion in the final orthomosaic map, which reduces accuracy and limits the potential for accurate measurements and reliable surveys. Moreover, most drones have been assigned with their operational limits when flying by themselves. In the case of customizations, the modifications alter the aircrafts aerodynamics, while the integration of additional payloads (i.e., the sensing devices) also affects the airborne response capacity and agility of the UAV. To address these issues, two common methods have been devised to ensure UAV stability: (1) the use of controllers that keep the aircraft level when hit by gusts, allowing it to recover quickly; and (2) the use of gimbals for the mounted cameras, to maintain the nadir (or generally predetermined) capture angle even when the UAV is not on level.

Fertilization management

Regarding input management, the use of fertilizers and fertilization management using UAVs is a topic that has received considerable attention by the scientific and manufacturing communities due to the ease with which field prescription maps are generated through UAV surveys (Avtar and Watanabe, 2020). UAVs have a variety of applications in agriculture, including but not being limited to fertilizer management, especially in complex heterogeneous agricultural environments and in various scenarios, where they can heavily assist the automation of tasks, such as fertilization, and significantly support decision making (Boursianis et al., 2020). Go et al. (2022) demonstrated that accurately predicting the growth stage of cabbage with the use of UAVs contributes greatly in the optimal management of fertilization schedules throughout the season. Moreover, the potential of UAV-derived VIs data to guide in-season fertilization timing decisions in growing carrots has been evaluated relative to conventional sampling methods, including petiole sap nitrate testing (Metiva, 2021). Optical sensors (e.g., Red Green Blue (RGB) or multispectral cameras) are used for remote sensing applications on UAVs for the purpose of monitoring nutrient status of vegetable crops and soils for optimal fertilizer (mainly nitrogen) management, in the context of determining the crop's needs using nondestructive methods (Padilla et al., 2020). The advantage of optical sensors is that the measurements are made instantly and the results are rapidly available.

In arable crops, nutrient stress detection can vary in difficulty, as there are many causal factors associated, while the fields are considerably larger in area and more homogeneous. As a result, several studies have focused on the identification of nitrogen and nutrient deficits. Furthermore, due to the larger area that arable fields cover, the autonomy of the deployed UAVs needs to be higher, while the resolution required is often significantly lower, for most applications, compared to vegetable fields. As a result, fixed-wing or vertical take-off and landing (VTOL) UAVs are often preferred for arable crop surveys (e.g., Maresma et al., 2016), as they can remain airborne for longer periods, and can thus more efficiently scan larger areas. Geipel et al. (2016) used VIs to predict both biomass and plant nitrogen content, while Tokekar et al. (2016) introduced a new approach using a mix of UAV and autonomous terrestrial vehicles, to determine nitrogen levels in the soil using reflectance indices. Vergara-Díaz et al. (2016) studied how NDVI data from a multispectral camera mounted on a UAV can be used to phenotype low-nitrogen stress tolerance in maize. The nitrogen balance marker (NBI), which analyzes the ratio of chlorophyll to polyphenols, is another index used to diagnose nitrogen deficit. Uses of this index have been mentioned by Gabriel et al. (2017) through their study on the ability of proximal and airborne sensors to identify the nutritional nitrogen status of maize. Maresma et al. (2016) used an ATMOS-6 UAV to acquire data from a maize crop divided into 45 experimental plots to analyze VIs to determine nitrogen application.

Disease and pest detection

Disease outbreaks and pest infestations are some of the most harmful factors for global agricultural production, as they are highly unpredictable and manifest unevenly within fields. Thus early detection and timely treatment applications are critical for their effective control and damage mitigation. At the same time, pesticides are one of the most important pollutants of the agricultural sector, as they are widely overused and end-up to water reserves, contributing to environmental degradation and severe human health consequences due to their high toxicity (Parron et al., 2014; Lampridi et al., 2019; Nuro, 2021). One of the major components of precision agriculture is site-specific pesticide management which can be performed with UAV monitoring. UAVs can be used in two phases of disease detection: early infection detection, before any visible signs manifest, and mapping of the degree of pathogen or pest infection (Maes and Steppe, 2019) after symptoms become visible. Crop diseases and insect infestations may be monitored with UAVs equipped with multispectral, hyperspectral, and thermal sensors. Due to physiological changes in cell structure, pigment, water content, leaf area index (LAI), and biomass produced by disease spots, wilting, defoliation, necrosis, poor growth, and other features, the spectrums of affected and healthy crops differ significantly (Zhang et al., 2021).

In several crops, the reflectance of infected plants in the visible light range is higher than healthy ones, while their reflectance in the NIR range is lower (Zhao et al., 2020), demonstrating the "blue shift" phenomenon. Diseases such as wheat stripe rust disease have been found using hyperspectral UAV photography and the photochemical vegetation index PRI (Huang et al., 2007). Wang et al. (2018) managed to accurately diagnose rice illness using a multirotor UAV system, reducing the impact of complicated background factors such as rice leaf occlusion, rice panicle adhesion, and lighting in the natural environment. Cui (2019) identified features of cotton field mites from remote sensing data that allow for real-time detection of mite damage. A study was conducted in Latin America using a UAV equipped with a multispectral and a thermal camera to phenotype tar spot complex resistance in maize. In this study, Loladze et al. (2019) observed a strong relationship firstly, between grain yield, a vegetation index (MCARI2) and canopy temperature under disease pressure, and secondly, between the area under the disease progress curve of tar spot complex (TSC) and three vegetative indices (RDVI, MCARI1, and MCARI2).

Weeds are among the most impacting biotic factors in agriculture (Oerke, 2006), causing important yield loss rates that exceed 50% for major crops such as corn and soybean worldwide (Esposito et al., 2021). An approach that minimizes most major drawbacks of chemical weed control while increasing the efficiency of weed-control applications is site-specific weed management. This can be done in two ways; either in real time using sensing systems mounted on spraying machinery, or through weed population mapping surveys and the generated occurrence maps. For the latter, among the wide range of technologies used for weed identification, detection, and mapping, UAVs stand out, since they enable efficient, precise, and continuous monitoring of large areas. Low altitude UAV imagery has been widely used in the area of weed detection to generate infestation maps to optimize weed control in various open-field crops, although they have been more extensively tested on arable crops such as wheat, winter barley, beet, and maize (Louargant et al., 2018; Franco et al., 2018; Fawakherji et al., 2019; Rasmussen et al., 2019), which are among the most cultivated crops worldwide and are highly susceptible to weed competition, especially in early phenological stages. In these crops, the detection of several weed types has been studied, including amaranth, lambsquarters, and Canada thistle (Hansen et al., 2013; Huang et al., 2018; Sanders et al., 2019). Rasmussen et al. (2013)

explored the potential of UAV-based weed detection in barley by analyzing UAV images and generating a quantitative weed map to facilitate weed control planning prior to the operation, helping to choose optimal herbicides, and spraying volume rates. Pérez-Ortiz et al. (2015, 2016) used UAVs and a crop row detection method for weed mapping in sunflower crops and maize. Peña et al. (2013) deployed a six-band multispectral camera to collect UAV images in an early-season maize field to map the position and coverage of the weeds, exporting the results in a customized grid structure adapted to a herbicide sprayer. Castaldi et al. (2017) used UAV imagery in the RGB and NIR spectral ranges to evaluate different postemergence herbicide application strategies in maize fields, reporting a range of herbicide savings between 14% and 39% as compared with a uniform application.

Spraying and pest management

UAVs are not only used for detection but also for treatment operations through aerial spraying technology. Multiple researches have also focused on improving crop management and input use for pest management spraying applications of insecticides and miticides through the use of spraying drones in vegetable crops (Iost Filho et al., 2020; Costa et al., 2012; Faiçal et al., 2014a). Kim et al. (2021) studied UAV spraying in cabbage fields and assessed the dissipation and distribution of chemical residues in the field. The use of UAVs for spraying in open-field vegetables, with the aim to increase the application precision efficiency and reduce negative impacts on humans and the environment is, also, proposed by the work of Faical et al. (2016). In this work, the use of UAVs able to self-adjust their route under difficult weather conditions is proposed (e.g., high wind speed). Similarly, Faiçal et al. (2014b) in their work talk about the ability to fine-tune UAV control rules and, ultimately, make more personalized decisions for different crop fields, in a more autonomous way, for the purpose of increasing the efficiency of UAV spraying, and also reducing spray drift in open field crops (Brown and Giles, 2018). Going even further in technological development, there is also a study that proposes that multiple UAV systems, using distributed swarm control algorithms, are superior to single UAV system for spraying applications in open-field crops (Ju and Son, 2018). Apart from UAV spraying for chemical plant protection products, there have been research activities focusing on organic pest management using drones, such as the distribution of lacewing eggs for aphid control in organic lettuce (Del Pozo-Valdivia et al., 2021).

In the field of spraying applications in arable crops, Wang et al. (2019) conducted research on the influence of spray volume on deposition and pest and disease control in wheat, concluding that with coarse nozzles, the UAV had comparable deposition and application efficacy to an electric air-pressure knapsack sprayer at a greater spraying volume rate (>16 L/ha), but had worse deposition and efficacy at a lower spraying rate (<9 L/ha). Additionally, the Chinese Ministry of Agriculture's Nanjing Research Institute for Agricultural Mechanization and China's National Center for International Collaboration Research on Precision Agricultural Aviation Pesticides Spraying Technology (NPAAC) investigated the relationship between different UAV working parameters and droplet deposition on rice (Chen et al., 2016; Xinyu et al., 2014) and maize (Qin et al., 2014), respectively.

Irrigation

Water availability in many agricultural areas across the world demonstrates high annual variability, often leading to periods when water is very scarce, resulting in drought stress in crops and dry soils which are highly vulnerable to nutrient loss. One of the primary challenges of the agricultural sector is to make optimum use of water resources (Rockström et al., 2017; Rodias et al., 2021). Precision irrigation and more specifically site-specific variable irrigation has been extensively studied as a solution to water shortages, as well as a method for minimizing salt and nutrient losses, and ensuring that adequate water is distributed in an optimal way (Ding, 2019).

Thermal imaging is a widely used remote sensing tool for assessing plant water stress, as plant water content is directly associated with their thermoregulation mechanisms (Michaletz et al., 2015). Using UAV-derived thermal imagery, crop-specific thermal indices have demonstrated considerable promise in determining field-level water stress heterogeneity (Gago et al., 2015). Hoffmann et al. (2016) conducted a study in two adjacent spring barley fields investigating whether the water deficit index (WDI) can be used to generate accurate crop water stress maps through images from a UAV. Results from this study showed that the UAV-based WDI maps were in agreement with ground truth water stress measurements, especially in the late growing stages. Another study for water-saving management, conducted at an experimental rice field of Taiwan Agriculture Research Institute (TARI), was based on rice crop height modeling with an RGB camera attached in an UAV and with in situ measurements, in order to perform variable rate irrigation based on crop height. The study revealed that the UAV height estimations had very high accuracy, indicating that the UAV images are able to determine rice's height and achieve major water usage reduction that ranged from 20% up to 50% (Yang et al., 2020). Another study which demonstrated the capacity of UAV-mounted thermal sensors in monitoring water content was conducted in Brazil for soybeans. In that study, Crusiol et al. (2020) used thermal imagery captured by a UAV thermal camera, to assess the water status of soybean plants exposed to various water availability conditions. The findings confirm that a water deficit during the reproductive phases of soybean growth has a greater impact on grain output than a deficit during the vegetative stages.

Yield prediction

Crop yield is the most important piece of information for crop management in precision agriculture. Early crop yield estimations are valuable insights that allow farmers to optimize farm-level decisions, like the upcoming farming operations scheduling or farm management decisions such as whether to sell or store the final product. At the same time, yield forecasts can act as an assessment tool for the expected income of each production unit, while also enabling an efficient transfer of the production from the farm to the food supply chain. Traditionally, crop yield estimations were performed by in-field sampling, such as destructive methods that include biomass weighting and grain size measurements during the latter growth stages of the crops. Naturally, these traditional methods are associated with numerous limitations such as the difficulty in applying manual sampling in large fields and the high level of uncertainty in the estimations due to the low volume of data that can be collected each season, thus often setting them as ineffective and unreliable.

The development of crop-specific models provides an accessible solution that can drastically decrease uncertainty while minimizing the efforts required for accurate yield forecasts. This approach essentially involves the identification of which parameters influence crop growth and development the most, and requires data from multiple years to create time series and yield data and recordings of several agrometeorological parameters to derive a regression equation that describes yield or maturity as a function of multiple agrometeorological parameters. Additionally, during the development phase, potential stress instances are identified and registered as risk factors which are also integrated into the models. This way, the models become robust and capable of providing accurate forecasts even in unusual situations, such as the occurrence of a heat stress or frost during certain growth stages that ultimately affect yield. Such models naturally become more accurate and robust as the volume and quality of data provided as inputs are increased (Zude-Sasse et al., 2021).

For approaches that use imagery data, the critical parameters are often distinguishable through visual characteristics of the crops. These factors include but are not limited to changes in spectral reflectance for cases where low resolution imagery data are used for field-level estimations, or crop-level identification, such as fruit detection, in cases where higher resolution data are used. Once these factors and their level of influence on crop growth have been determined through experimental trials, the weight of each factor is assigned to the respective parameter in the model's mathematical equations. As a result, in the following growing seasons, these factors are then used as inputs in the model generating estimations for crop growth, development, and ultimately yield based on data from the current season (Hoogenboom et al., 2004). In the case of remote sensing-based models, field-level measurements collected in the experimental sites are used as reference data for validation of the remote imagery, as well as a calibration factor of the models. In case a crop-type demonstrates rapid change in its characteristics, the correlation of a certain parameter with final yield is considered significant only after a certain growth stage (i.e., the flowering of *Brassicaceae* plants), with this period being identified as a critical stage, and data during this period are assigned a higher weight due to their direct correlation with maturity level and final yield.

Another method of yield estimation involves the use of photogrammetry, for the three-dimensional (3D) representation of crops. This process aims to reconstruct a 3D model of the surveyed area, and in agricultural fields, the same principles apply for the data collection using imaging sensors (i.e., cameras), where point clouds are generated from aerial images through photogrammetric algorithms, for example, structure from motion—SfM (Ullman, 1979). However, sensing systems such as LiDAR and ultrasound sensors are also used to collect georeferenced 3D data with higher accuracy, without being heavily affected by illumination.

Moreover, sensor fusion approaches that essentially combine data from different systems are used to higher accuracy models (Grüner et al., 2021). These approaches have been widely used for the estimation of above ground biomass through the calculation of crop height or dimensions, which for many crops is directly connected to yield. UAVs equipped with commercial RGB cameras and simple photogrammetric implementations have been successfully used for the estimation of biomass in both arable crops (e.g., Bendig et al., 2014; Holman et al., 2016; Schirrmann et al., 2016) and vegetables (Moeckel et al., 2018). Moreover, in vegetable production, low altitude UAV imagery enables plant-level monitoring, retrieving information regarding the condition of individual plants. In cases such as most open-field vegetables, where the harvested part of the plant is visible from above, UAV images can be used to classify them and predict both their maturity level and estimated yield. Psiroukis et al. (2022) used low altitude (10 m) RGB UAV orthomosaic-derived images to detect broccoli crops and classify them based on their maturity level in three maturity classes though object detection. Kim et al. (2018) used UAV-captured RGB images from a slightly higher altitude (20 m) to identify spatial and temporal patterns in biophysical properties of cabbage and radish fields.

UAVs are also highly valuable for providing information on within-field variation of growth stage and biomass in arable crops, which are highly homogeneous, and underlying variability in vigor is mainly connected to growth stage or inherent soil variations, in contrast to vegetables, which are inherently more susceptible to micro-climatic variations. Although object detection tasks cannot be deployed for yield estimation in arable crops, a variety of VIs has been used to monitor crop status and quantitatively evaluate the vitality of the fields. Marino and Alvino (2020) used the soil-adjusted vegetation index (SAVI), NDVI, and OSAVI to characterize 10 winter wheat varieties in a field at different growth stages, obtaining optimal biomass monitoring results. The growth stages of cereal crops can be determined in great detail with VIs derived from UAV-based RGB imagery (Schirrmann et al., 2016; Du and Noguchi, 2017; Burkart et al., 2018). Biophysical variables such as canopy coverage (LAI) and FAPAR (fraction of absorbed photosynthetically active radiation) are also used for determining spatial variation within the fields (Duan et al., 2014; Verger et al., 2014). Apart from VIs, biomass is also strongly correlated with crop height and final yield (Montes et al., 2011; Hakl et al., 2012). Therefore several studies in arable crops have been focused on estimating vegetation height based on UAV (RGB) imagery, providing promising results (Madec et al., 2017; Watanabe et al., 2017; Hu et al., 2018; Brocks and Bareth, 2018). Characteristically, Hassan et al. (2019) used a multispectral camera to study the accuracy of UAV-derived height estimations for wheat crops, observing a high correlation between the crop height data from UAV and the actual height, setting UAV photogrammetry as a reliable tool for wheat yield estimation. Accurate estimates of biomass were also obtained by combining vegetation height estimates with one or more VIs from multispectral data (Bendig et al., 2015; Yue et al., 2017). With UAVs, high yield prediction accuracies have also been recorded by applying RGB-derived plant height and canopy cover (Chu et al., 2016), VIs (Wahab et al., 2018, Gracia-Romero et al., 2017; Du and Noguchi, 2017), and multispectral indices (Kyratzis et al., 2017; Zhou et al., 2017).

Conclusions

The objective of this chapter was to cover the necessary fundamental principles on operational surveying aspects of UAVs, in order to follow up with an extended overview of available applications and active research carried out in the vegetable and arable farming sectors. As these crops share several similarities, but also differences, the cases are divided appropriately, to provide context for the reader, regardless of their knowledge and background. The replication of these applications, however, in each specific scenario, is a highly challenging task. In the selection of the UAV alone, multiple different characteristics should be considered, such as the system's performance, autonomy, and load capacity, which are all critical for agricultural applications. To this end, a detailed analysis should always be conducted, taking into consideration the unique aspects and requirements of each foreseen application, and then deciding on the optimal UAV and sensor configuration to achieve these objectives. Moreover, a proper risk management plan should be included in the analysis, as environmental factors such as turbulence, gusts, and extreme temperatures drastically affect the components, performance, and safety of the aircraft and nearby people. Furthermore, challenges regarding autonomy and flight duration, aircraft stability, susceptibility to weather conditions, regulations and restrictions regarding their operation, platform and sensing components' failures, and payload capacity are all active problems that are being currently addressed by the UAVs manufacturers. Therefore careful consideration of both proper aircraft components and mission planning is the only way to effectively address the aforementioned factors, which is crucial for agricultural UAVs applications (Katikaridis et al., 2022). Finally, concerns about operation safety and flight regulations emerged as the number of UAVs operating across the globe, for both recreational and professional purposes, are increasing rapidly. The number of UAVs and UAV operators in rural areas are expected to constantly continue growing, considering the multiple advantages and solutions to active problems that this new technology offers to farmers. At the same time, the decrease in purchasing costs, for both the aircrafts and the sensing components, will greatly assist in achieving greater adoption rates across the world, making UAV solutions accessible to both smallholders and larger farmers alike.

References

Avtar, R., Watanabe, T., 2020. Unmanned Aerial Vehicle: Applications in Agriculture and Environment. Springer Nature Switzerland AGISBN 978-3-030-27157-2 (eBook). Available from: https://doi.org/10.1007/978-3-030-27157-2.

Bendig, J., Bolten, A., Bennertz, S., Broscheit, J., Eichfuss, S., Bareth, G., 2014. Estimating biomass of barley using crop surface models (CSMs) derived from UAV-based RGB imaging. Remote Sens. 6 (11), 10395−10412.

Bendig, J., Yu, K., Aasen, H., Bolten, A., Bennertz, S., Broscheit, J., et al., 2015. Combining UAV-based plant height from crop surface models, visible, and near infrared vegetation indices for biomass monitoring in barley. Int. J. Appl. Earth Obs. Geoinf. 39, 79−87.

Berni, J., Zarco-Tejada, P.J., Suarez, L., Fereres, E., 2009. Thermal and narrowband multispectral remote sensing for vegetation monitoring from an unmanned aerial vehicle. IEEE Trans. Geosci. Remote Sens. 47, 722−738.

Boursianis, A.D., Papadopoulou, M.S., Diamantoulakis, P., Liopa-Tsakalidi, A., Barouchas, P., Salahas, G., et al., 2020. Internet of things (IoT) and agricultural unmanned aerial vehicles (UAVs) in smart farming: a comprehensive review. Internet Things 100187.

Brocks, S., Bareth, G., 2018. Estimating barley biomass with crop surface models from oblique RGB imagery. Remote Sens. 10 (2), 268.

Brown, C.R., Giles, D.K., 2018. Measurement of pesticide drift from unmanned aerial vehicle application to a vineyard. Trans. ASABE 61 (5), 1539−1546.

Burkart, A., Aasen, H., Alonso, L., Menz, G., Bareth, G., Rascher, U., 2015. Angular dependency of hyperspectral measurements over wheat characterized by a novel UAV based goniometer. Remote Sens. 7, 725−746.

Burkart, A., Hecht, V.L., Kraska, T., Rascher, U., 2018. Phenological analysis of unmanned aerial vehicle based time series of barley imagery with high temporal resolution. Precis. Agric. 19 (1), 134−146.

Castaldi, F., Pelosi, F., Pascucci, S., Casa, R., 2017. Assessing the potential of images from unmanned aerial vehicles (UAV) to support herbicide patch spraying in maize. Precis. Agric. 18 (1), 76−94.

Chen, S., Lan, Y., Li, J., Zhou, Z., Jin, J., Liu, A., 2016. Effect of spray parameters of small unmanned helicopter on distribution regularity of droplet deposition in hybrid rice canopy. Trans. Chin. Soc. Agric. Eng. 32 (17), 40−46.

Chu, T., Chen, R., Landivar, J.A., Maeda, M.M., Yang, C., Starek, M.J., 2016. Cotton growth modeling and assessment using unmanned aircraft system visual-band imagery. J. Appl. Remote Sens. 10 (3), 036018.

Cihlar, J., Laurent, L.S., Dyer, J.A., 1991. Relation between the normalized difference vegetation index and ecological variables. Remote Sens. Environ. 35 (2−3), 279−298.

Costa, F.G., Ueyama, J., Braun, T., Pessin, G., Osório, F.S., Vargas, P.A., 2012. The use of unmanned aerial vehicles and wireless sensor network in agricultural applications, in: 2012 IEEE International Geoscience and Remote Sensing Symposium (pp. 5045-5048).

Crusiol, L.G.T., Nanni, M.R., Furlanetto, R.H., Sibaldelli, R.N.R., Cezar, E., Mertz-Henning, L.M., et al., 2020. UAV-based thermal imaging in the assessment of water status of soybean plants. Int. J. Remote Sens. 41 (9), 3243−3265.

Cui, M.N., 2019 Study on Dynamic Monitoring of Cotton Spider Mites Based on Remote Sensing of UAV. Master's Thesis, Shihezi University, Shihezi, China.

Ding, K., 2019. State of knowledge of irrigation techniques and practicalities within given socio-economic settings. Irrig. Drain. 68 (1), 31−45.

Duan, S.B., Li, Z.L., Wu, H., Tang, B.H., Ma, L., Zhao, E., et al., 2014. Inversion of the PROSAIL model to estimate leaf area index of maize, potato, and sunflower fields from unmanned aerial vehicle hyperspectral data. Int. J. Appl. Earth Obs. Geoinf. 26, 12−20.

Du, M., Noguchi, N., 2017. Monitoring of wheat growth status and mapping of wheat yield's within-field spatial variations using color images acquired from UAV-camera system. Remote Sens. 9 (3), 289.

Esposito, M., Crimaldi, M., Cirillo, V., Sarghini, F., Maggio, A., 2021. Drone and sensor technology for sustainable weed management: a review. Chem. Biol. Technol. Agric. 8 (1), 1−11.

Faical, B.S., Ueyama, J., de Carvalho, A.C., 2016. The use of autonomous UAVs to improve pesticide application in crop fields, in: 2016 17th IEEE International Conference on Mobile Data Management (MDM), vol. 2, 32−33. IEEE.

Faiçal, B.S., Costa, F.G., Pessin, G., Ueyama, J., Freitas, H., Colombo, A., et al., 2014a. The use of unmanned aerial vehicles and wireless sensor networks for spraying pesticides. J. Syst. Architect. 60, 393−404.

Faiçal, B.S., Pessin, G., Filho, G.P.R., Carvalho, A.C.P.L.F., Furquim, G., Ueyama, J., 2014b. Fine-tuning of UAV control rules for spraying pesticides on crop fields, in: IEEE 26th International Conference on Tools with Artificial Intelligence, 527−533, Available from: https://doi.org/10.1109/ICTAI.2014.85.

Fawakherji, M., Potena, C., Bloisi, D.D., Imperoli, M., Pretto, A., Nardi, D., 2019. UAV image-based crop and weed distribution estimation on embedded GPU boards, in: International Conference on Computer Analysis of Images and Patterns, 100−108.

Franco, C., Guada, C., Rodríguez, J.T., Nielsen, J., Rasmussen, J., Gómez, D., et al., 2018. Automatic detection of thistle-weeds in cereal crops from aerial RGB images, in: International Conference on Information Processing and Management of Uncertainty in Knowledge-Based Systems, 441−452.

Gabriel, J.L., Zarco-Tejada, P.J., López-Herrera, P.J., Pérez-Martín, E., Alonso-Ayuso, M., Quemada, M., 2017. Airborne and ground level sensors for monitoring nitrogen status in a maize crop. Biosyst. Eng. 160, 124−133.

Gago, J., Douthe, C., Coopman, R.E., Gallego, P.P., Ribas-Carbo, M., Flexas, J., et al., 2015. UAVs challenge to assess water stress for sustainable agriculture. Agric. Water Manag. 153, 9−19.

Geipel, J., Link, J., Wirwahn, J.A., Claupein, W., 2016. A programmable aerial multispectral camera system for in-season crop biomass and nitrogen content estimation. Agriculture 6 (1), 4.

Go, S.H., Lee, D.H., Na, S.I., Park, J.H., 2022. Analysis of growth characteristics of kimchi cabbage using drone-based cabbage surface model image. Agriculture 12 (2), 216.

Gracia-Romero, A., Kefauver, S.C., Vergara-Diaz, O., Zaman-Allah, M.A., Prasanna, B. M., Cairns, J.E., et al., 2017. Comparative performance of ground vs. aerially assessed RGB and multispectral indices for early-growth evaluation of maize performance under phosphorus fertilization. Front. Plant Sci. 8, 2004.

Grüner, E., Astor, T., Wachendorf, M., 2021. Prediction of biomass and N fixation of legume−grass mixtures using sensor fusion. Front. Plant Sci. 2192.

Haala, N., Rothermel, M., 2012. Dense multiple stereo matching of highly overlapping UAV imagery. ISPRS J. Photogramm. Remote Sens.

Hakl, J., Hrevušová, Z., Hejcman, M., Fuksa, P., 2012. The use of a rising plate meter to evaluate lucerne (Medicago sativa L.) height as an important agronomic trait enabling yield estimation. Grass Forage Sci. 67 (4), 589−596.

Hansen, K.D., Garcia-Ruiz, F., Kazmi, W., Bisgaard, M., la Cour-Harbo, A., Rasmussen, J., et al., 2013. An autonomous robotic system for mapping weeds in fields. IFAC Proc. Volumes 46 (10), 217−224.

Hassan, M.A., Yang, M., Fu, L., Rasheed, A., Zheng, B., Xia, X., et al., 2019. Accuracy assessment of plant height using an unmanned aerial vehicle for quantitative genomic analysis in bread wheat. Plant Methods 15 (1), 1−12.

Hoffmann, H., Jensen, R., Thomsen, A., Nieto, H., Rasmussen, J., Friborg, T., 2016. Crop water stress maps for an entire growing season from visible and thermal UAV imagery. Biogeosciences 13 (24), 6545−6563.

Holman, F.H., Riche, A.B., Michalski, A., Castle, M., Wooster, M.J., Hawkesford, M.J., 2016. High throughput field phenotyping of wheat plant height and growth rate in field plot trials using UAV based remote sensing. Remote Sens. 8 (12), 1031.

Hoogenboom, G., White, J.W., Messina, C.D., 2004. From genome to crop: integration through simulation modeling. Field Crop. Res. 90 (1), 145−163.

Huang, W., Lamb, D.W., Niu, Z., Zhang, Y., Liu, L., Wang, J., 2007. Identification of yellow rust in wheat using in-situ spectral reflectance measurements and airborne hyperspectral imaging. Precis. Agric. 8 (4), 187−197.

Huang, Y., Reddy, K.N., Fletcher, R.S., Pennington, D., 2018. UAV low-altitude remote sensing for precision weed management. Weed Technol. 32 (1), 2−6.

Hu, P., Chapman, S.C., Wang, X., Potgieter, A., Duan, T., Jordan, D., et al., 2018. Estimation of plant height using a high throughput phenotyping platform based on unmanned aerial vehicle and self-calibration: example for sorghum breeding. Eur. J. Agron. 95, 24−32.

Iost Filho, F.H., Heldens, W.B., Kong, Z., de Lange, E.S., 2020. Drones: innovative technology for use in precision pest management. J. Econ. Entomol. 113 (1), 1−25.

Ju, C., Son, H.I., 2018. Multiple UAV systems for agricultural applications: control, implementation, and evaluation. Electronics 7 (9), 162.

Katikaridis, D., Moysiadis, V., Tsolakis, N., Busato, P., Kateris, D., Pearson, S., et al., 2022. UAV-supported route planning for UGVs in semi-deterministic agricultural environments. Agronomy 12 (8), 1937.

Kim, C.J., Jeong, W.T., Kyung, K.S., Lee, H.D., Kim, D., Song, H.S., et al., 2021. Dissipation and distribution of picarbutrazox residue following spraying with an unmanned aerial vehicle on Chinese cabbage (Brassica campestris var. pekinensis). Molecules 26 (18), 5671.

Kim, D.W., Yun, H.S., Jeong, S.J., Kwon, Y.S., Kim, S.G., Lee, W.S., et al., 2018. Modeling and testing of growth status for Chinese cabbage and white radish with UAV-based RGB imagery. Remote Sens. 10 (4), 563.

Kyratzis, A.C., Skarlatos, D.P., Menexes, G.C., Vamvakousis, V.F., Katsiotis, A., 2017. Assessment of vegetation indices derived by UAV imagery for durum wheat phenotyping under a water limited and heat stressed mediterranean environment. Front. Plant Sci. 8, 1114.

Lampridi, M.G., Sørensen, C.G., Bochtis, D., 2019. Agricultural sustainability: a review of concepts and methods. Sustainability 11 (18), 5120.

Loladze, A., Rodrigues Jr, F.A., Toledo, F., San Vicente, F., Gérard, B., Boddupalli, M.P., 2019. Application of remote sensing for phenotyping tar spot complex resistance in maize. Front. Plant Sci. 10, 552.

Louargant, M., Jones, G., Faroux, R., Paoli, J.N., Maillot, T., Gée, C., et al., 2018. Unsupervised classification algorithm for early weed detection in row-crops by combining spatial and spectral information. Remote Sens. 10 (5), 761.

Madec, S., Baret, F., De Solan, B., Thomas, S., Dutartre, D., Jezequel, S., et al., 2017. High-throughput phenotyping of plant height: comparing unmanned aerial vehicles and ground LiDAR estimates. Front. Plant Sci. 8, 2002.

Maes, W.H., Steppe, K., 2019. Perspectives for remote sensing with unmanned aerial vehicles in precision agriculture. Trends Plant Sci. 24 (2), 152–164.

Maresma, Á., Ariza, M., Martínez, E., Lloveras, J., Martínez-Casasnovas, J.A., 2016. Analysis of vegetation indices to determine nitrogen application and yield prediction in maize (Zea mays L.) from a standard UAV service. Remote Sens. 8 (12), 973.

Marino, S., Alvino, A., 2020. Agronomic traits analysis of ten winter wheat cultivars clustered by UAV-derived vegetation indices. Remote Sens. 12 (2), 249.

Matese, A., Toscano, P., Di Gennaro, S.F., Genesio, L., Vaccari, F.P., Primicerio, J., et al., 2015. Intercomparison of UAV, aircraft and satellite remote sensing platforms for precision viticulture. Remote Sens. 7, 2971–2990. Available from: https://doi.org/10.3390/rs70302971.

Metiva, M.A., 2021. Applications of Drone-Based Remote Sensing in Carrot and Tomato Cropping Systems. Michigan State University.

Michaletz, S.T., Weiser, M.D., Zhou, J., Kaspari, M., Helliker, B.R., Enquist, B.J., 2015. Plant thermoregulation: energetics, trait–environment interactions, and carbon economics. Trends Ecol. Evol. 30 (12), 714–724.

Moeckel, T., Dayananda, S., Nidamanuri, R.R., Nautiyal, S., Hanumaiah, N., Buerkert, A., et al., 2018. Estimation of vegetable crop parameter by multi-temporal UAV-borne images. Remote Sens. 10 (5), 805.

Mondal, P., Basu, M., 2009. Adoption of precision agriculture technologies in India and in some developing countries: scope, present status and strategies. Prog. Nat. Sci. 19 (6), 659–666.

Montes, J.M., Technow, F., Dhillon, B.S., Mauch, F., Melchinger, A.E., 2011. High-throughput non-destructive biomass determination during early plant development in maize under field conditions. Field Crop Res. 121 (2), 268–273.

Moran, M.S., Inoue, Y., Barnes, E.M., 1997. Opportunities and limitations for image-based remote sensing in precision crop management. Remote Sens. Environ. 61 (3), 319–346.

Ni, W., Sun, G., Pang, Y., Zhang, Z., Liu, J., Yang, A., et al., 2018. Mapping three-dimensional structures of forest canopy using UAV stereo imagery: evaluating impacts of forward overlaps and image resolutions with LiDAR data as reference. IEEE J. Sel. Top. Appl. Earth Obs. Remote Sens. 11 (10), 3578–3589.

Nuro, A. (Ed.), 2021. Emerging Contaminants. IntechOpen. Available from: https://doi.org/10.5772/intechopen.87857.

Oerke, E.C., 2006. Crop losses to pests. J. Agric. Sci. 144 (1), 31–43.

Padilla, F.M., Farneselli, M., Gianquinto, G., Tei, F., Thompson, R.B., 2020. Monitoring nitrogen status of vegetable crops and soils for optimal nitrogen management. Agric. Water Manag. 241, 106356.

Parron, T., Requena, M., Hernández, A.F., Alarcón, R., 2014. Environmental exposure to pesticides and cancer risk in multiple human organ systems. Toxicol. Lett. 230 (2), 157–165.

Peña, J.M., Torres-Sánchez, J., de Castro, A.I., Kelly, M., López-Granados, F., 2013. Weed mapping in early-season maize fields using object-based analysis of unmanned aerial vehicle (UAV) images. PLoS One 8 (10), e77151.

Del Pozo-Valdivia, A.I., Morgan, E., Bennett, C., 2021. In-field evaluation of drone-released lacewings for aphid control in California organic lettuce. J. Econ. Entomol. 114 (5), 1882–1888.

Psiroukis, V., Espejo-Garcia, B., Chitos, A., Dedousis, A., Karantzalos, K., Fountas, S., 2022. Assessment of different object detectors for the maturity level classification of broccoli crops using UAV imagery. Remote Sens. 14 (3), 731.

Psiroukis, V., Malounas, I., Mylonas, N., Grivakis, K.E., Fountas, S., Hadjigeorgiou, I., 2021. Monitoring of free-range rabbits using aerial thermal imaging. Smart Agric. Technol. 1, 100002.

Pérez-Ortiz, M., Peña, J.M., Gutiérrez, P.A., Torres-Sánchez, J., Hervás-Martínez, C., López-Granados, F., 2015. A semi-supervised system for weed mapping in sunflower crops using unmanned aerial vehicles and a crop row detection method. Appl. Soft Comput. 37, 533–544.

Pérez-Ortiz, M., Peña, J.M., Gutiérrez, P.A., Torres-Sánchez, J., Hervás-Martínez, C., López-Granados, F., 2016. Selecting patterns and features for between-and within-crop-row weed mapping using UAV-imagery. Expert. Syst. Appl. 47, 85–94.

Qin, W., Xue, X., Zhou, L., Zhang, S., Sun, Z., Kong, W., et al., 2014. Effects of spraying parameters of unmanned aerial vehicle on droplets deposition distribution of maize canopies. Trans. Chin. Soc. Agric. Eng. 30 (5), 50–56.

Rasmussen, J., Nielsen, J., Garcia-Ruiz, F., Christensen, S., Streibig, J.C., 2013. Potential uses of small unmanned aircraft systems (UAS) in weed research. Weed Res. 53 (4), 242–248.

Rasmussen, J., Nielsen, J., Streibig, J.C., Jensen, J.E., Pedersen, K.S., Olsen, S.I., 2019. Pre-harvest weed mapping of Cirsium arvense in wheat and barley with off-the-shelf UAVs. Precis. Agric. 20 (5), 983–999.

Rockström, J., Williams, J., Daily, G., Noble, A., Matthews, N., Gordon, L., et al., 2017. Sustainable intensification of agriculture for human prosperity and global sustainability. Ambio 46 (1), 4–17.

Rodias, E., Aivazidou, E., Achillas, C., Aidonis, D., Bochtis, D., 2021. Water-energy-nutrients synergies in the agrifood sector: a circular economy framework. Energies 14 (1), 159. Available from: https://doi.org/10.3390/en14010159.

Sanders, J.T., Everman, W.J., Austin, R., Roberson, G.T., Richardson, R.J., 2019. Weed species differentiation using spectral reflectance land image classification, in: Advanced Environmental, Chemical, and Biological Sensing Technologies XV, 11007, 110070P. International Society for Optics and Photonics.

Schirrmann, M., Giebel, A., Gleiniger, F., Pflanz, M., Lentschke, J., Dammer, K.H., 2016. Monitoring agronomic parameters of winter wheat crops with low-cost UAV imagery. Remote Sens. 8 (9), 706.

Tokekar, P., Vander Hook, J., Mulla, D., Isler, V., 2016. Sensor planning for a symbiotic UAV and UGV system for precision agriculture. IEEE Trans. Robot. 32 (6), 1498–1511.

Tsouros, D.C., Bibi, S., Sarigiannidis, P.G., 2019. A review on UAV-based applications for precision agriculture. Information 10, 349. Available from: https://doi.org/10.3390/info10110349.

Tucker, C.J., 1979. Red and photographic infrared linear combinations for monitoring vegetation. Remote Sens. Environ. 8 (2), 127–150.

Tucker, C.J., 1980. Remote sensing of leaf water content in the near infrared. Remote Sens. Environ. 10 (1), 23—32.

Ullman, S., 1979. The interpretation of structure from motion. Proc. R. Soc. London. Ser. B. Biol. Sci. 203 (1153), 405—426.

Vergara-Díaz, O., Zaman-Allah, M.A., Masuka, B., Hornero, A., Zarco-Tejada, P., Prasanna, B.M., et al., 2016. A novel remote sensing approach for prediction of maize yield under different conditions of nitrogen fertilization. Front. Plant Sci. 7, 666.

Verger, A., Vigneau, N., Chéron, C., Gilliot, J.M., Comar, A., Baret, F., 2014. Green area index from an unmanned aerial system over wheat and rapeseed crops. Remote Sens. Environ. 152, 654—664.

Wahab, I., Hall, O., Jirström, M., 2018. Remote sensing of yields: application of uav imagery-derived ndvi for estimating maize vigor and yields in complex farming systems in sub-Saharan africa. Drones 2 (3), 28.

Wang, Z., Chu, G.K., Zhang, H.J., Liu, S.X., Huang, X.C., Gao, F.R., et al., 2018. Identification of diseased empty rice panicles based on Haar-like feature of UAV optical image. Trans. CSAE 34 (20), 73—82.

Wang, G., Lan, Y., Qi, H., Chen, P., Hewitt, A., Han, Y., 2019. Field evaluation of an unmanned aerial vehicle (UAV) sprayer: effect of spray volume on deposition and the control of pests and disease in wheat. Pest. Manag. Sci. 75 (6), 1546—1555.

Watanabe, K., Guo, W., Arai, K., Takanashi, H., Kajiya-Kanegae, H., Kobayashi, M., et al., 2017. High-throughput phenotyping of sorghum plant height using an unmanned aerial vehicle and its application to genomic prediction modeling. Front. Plant Sci. 8, 421.

Xing, C., Wang, J., Xu, Y., 2010. Overlap analysis of the images from unmanned aerial vehicles, in: 2010 International Conference on Electrical and Control Engineering. Available from: https://doi.org/10.1109/icece.2010.360.

Xinyu, X., Kang, T., Weicai, Q., Yubin, L., Huihui, Z., 2014. Drift and deposition of ultra-low altitude and low volume application in paddy field. Int. J. Agric. Biol. Eng. 7 (4), 23—28.

Yang, G., Liu, J., Zhao, C., Li, Z., Huang, Y., Yu, H., et al., 2017. Unmanned aerial vehicle remote sensing for field-based crop phenotyping: current status and perspectives. Front. Plant Sci. 8, 1111.

Yang, C.Y., Yang, M.D., Tseng, W.C., Hsu, Y.C., Li, G.S., Lai, M.H., et al., 2020. Assessment of rice developmental stage using time series UAV imagery for variable irrigation management. Sensors 20 (18), 5354.

Yue, J., Yang, G., Li, C., Li, Z., Wang, Y., Feng, H., et al., 2017. Estimation of winter wheat above ground biomass using unmanned aerial vehicle-based snapshot hyperspectral sensor and crop height improved models. Remote Sens. 9 (7), 708.

Zhang, H., Wang, L., Tian, T., Yin, J., 2021. A review of unmanned aerial vehicle low-altitude remote sensing (UAV-LARS) use in agricultural monitoring in China. Remote Sens. 13 (6), 1221.

Zhao, H., Yang, C., Guo, W., Zhang, L., Zhang, D., 2020. Automatic estimation of crop disease severity levels based on vegetation index normalization. Remote Sens. 12 (12), 1930.

Zhou, X., Zheng, H.B., Xu, X.Q., He, J.Y., Ge, X.K., Yao, X., et al., 2017. Predicting grain yield in rice using multi-temporal vegetation indices from UAV-based multi-spectral and digital imagery. ISPRS J. Photogramm. Remote Sens. 130, 246—255.

Zude-Sasse, M., Akbari, E., Tsoulias, N., Psiroukis, V., Fountas, S., Ehsani, R., 2021. Sensing in precision horticulture. Sens. Approaches Precis. Agric. 221—251.

CHAPTER 5

Unmanned aerial systems applications in orchards and vineyards

Aikaterini Kasimati, Ari Lomis, Vasilis Psiroukis, Nikoleta Darra, Michael Gerasimos Koutsiaras, George Papadopoulos and Spyros Fountas
Department of Natural Resources Management and Agricultural Engineering, Agricultural University of Athens, Athens, Greece

Introduction

Fruit tree growth involves a variety of labor-intensive and time-consuming monitoring and management tasks, such as pruning, watering, fertilizing, spraying, and harvesting. Automation in fruit tree management is important in precision agriculture because it not only reduces labor and increases grower income, but also improves resource utilization. Recent technological advances have made unmanned aerial vehicles (UAVs), also known as unmanned aerial systems (UAS) or "drones," an effective monitoring tool for improving orchard management, providing growers with much more detailed and precise information on fruit plant health, geometric variables, physiological variables, among others (Zhang et al., 2019).

Increased fruit production is a cost-effective solution to meet limited orchard monitoring needs, such as assessing growth and nutrition (Johansen et al., 2018). Farmers benefit from automation and precision orchard management by not only increasing yields but also reducing their environmental impact. Orchard management under precision agriculture requires accurate real-time monitoring of productivity, trees health, and water stress. However, these data sets are often difficult to obtain and, in most cases, come at a significant cost.

The present wave of studies onto the use of drones in orchard management has yielded new knowledge. However, since there are no standard operating procedures for employing drones in orchard and vineyard management, these studies are impediments to determining optimal techniques. A number of management activities are carried out throughout

Unmanned Aerial Systems in Agriculture
DOI: https://doi.org/10.1016/B978-0-323-91940-1.00005-0

the growth cycle of fruit crops. Different activities focus on managing fruit trees at various phases of development, and some activities are not confined to a particular stage. The growing state of tree crops can be monitored using changes in structural properties of fruit trees, and site-specific fungicide and water treatments can be applied.

The physical structure of fruit trees, such as the volume of the canopy that intercepts light, affects fruit quality and productivity (Rom, 1991). The most important approach to evaluating the effects of pruning is to consider geometric parameters such as the canopy area. Under deficit irrigation, orchard water delivery should match the practical needs of specific fruit crops. Mild water stress is commonly used in orchards to maintain or improve fruit quality and productivity. Therefore proper monitoring is needed to optimize the use of water resources.

Drones are increasingly available to address the demand for rapid and real-time monitoring times for orchard management with useful spatial, spectral, and temporal resolutions, thanks to developments in sensors technology that have lowered the overall size of sensors (Berni et al., 2009a; Torres-Sanchez et al., 2018a, 2018b; Caruso et al., 2019; Valente et al., 2019). They are frequently employed to transport remote sensing equipment due to their flexibility in flight planning, agility, and low-cost management. Data collection options for direct monitoring of fruit trees include light detection and ranging (LiDAR) devices, multispectral sensors as well as thermal sensors (Zhang et al., 2021). When choosing a UAV to carry these sensors and attain the needed time frequency and spatial resolution for orchard management, there are various factors to consider.

In the followings, we will explore UAV applications in:
- UAV-based monitoring, sensing, and mapping applications.
- Pest control and drone spraying.
- Irrigation efficiency.
- Yield prediction and selective harvesting.
- Flight planning for orchards and vineyards.

Monitoring, sensing, and mapping applications

UAV technology offers a faster, safer, and less expensive way to collect large amounts of data, and thus has the potential to transform agricultural monitoring, mapping, and application techniques. UAVs give researchers an unprecedented opportunity to track the evolution and dynamics of tree growth and structure over time. Hassler and Baysal-Gurel (2019) provided an overview of the state of the art in remote sensing with UAVs, as well as some application examples. As plant surface temperature fluctuates rapidly under stress conditions, the potential of UAV thermal remote sensing for mapping and monitoring lends itself to yield prediction, plant phenotyping (Ludovici et al., 2017; Kasimati et al., 2021a), water stress detection (Gago et al., 2015), and plant disease identification (Calderón et al., 2013), among other applications. Generation of vegetation index maps (Navia et al., 2016), such as normalized vegetation index (NDVI) maps, measurement of spatial variability and structure of vineyards (Nolan et al., 2015, Kasimati et al., 2021b), survey and creation of land cover maps (Murugan et al., 2017), and estimation of tree locations and sizes (Surový et al., 2018), are some notable studies using field mapping in agriculture. These maps are often created by stitching together a series of independent images taken by a UAS using various photogrammetric software programs (Roth and Streit, 2017). Ground control points (GCPs) are typically used to geo-reference each individual image in the resulting mosaic. Finally, Yuan and Choi (2021) aimed to demonstrate the concept of high-resolution, real-time UAV-based assessment of heating requirements for frost protection in apple orchards. They demonstrated how drones with thermal and RGB (red—green—blue) cameras can be used to assess heating needs for frost protection (Fig. 5.1).

The geometric characteristics of trees, such as shape and size, have been shown to be crucial clues for a number of activities in fruit tree management. By mapping these characteristics, it is possible to select and apply the best pruning method and intensity (Castillo-Ruiz et al., 2015; Miranda-Fuentes et al., 2015). In general, structural factors can be examined to identify tree crop growth status. Plant height measurement is a typical UAV application, as structure-from-motion (Zarco-Tejada et al., 2014; Jiménez-Brenes et al., 2017) can be used to create photogrammetrically derived digital surface models (DSMs) from overlapping images with multiple viewpoints of the same feature. Plant height can be used to model biomass, which is important information for predicting agricultural production (Matese et al., 2017). In addition, these geometric features can be used to design site-specific water treatments and solve management problems caused by soil heterogeneity.

The use of multispectral imagery can support in investigating the effects of pruning on changes in tree structure in commercial orchards. It also highlights the importance of capturing images with constant flight configurations, as a change in flight altitude can affect tree structure measurements. Pruning trees is thought to promote new growth, improve

Figure 5.1 A Phantom 4 Pro unmanned aerial vehicle in use within the vineyard to collect proximal imagery and a ground control point target designed to be easily detected in thermal as well as in multispectral unmanned aerial vehicle surveys.

fruiting, facilitate fruit harvest, and potentially increase production by increasing light absorption and canopy surface area. Johansen et al. (2018) used multispectral UAV imagery taken before and after pruning in a commercial lychee orchard to investigate a novel approach to measuring changes in tree structure such as canopy circumference, width, height, area, and projective plant cover.

Recent research has focused on automated 3D reconstruction techniques (3D digital models) for quantifying geometric features of fruit crops using drones (Ok and Ozdarici-Ok, 2018; Tagarakis et al., 2022). A well-established method for detecting fruit trees and determining geometric indices, such as perimeter, crown width, and tree height, is the combination of geographic object-based image analysis (GEOBIA) with DSM (Torres-Sánchez et al., 2015; Jiménez-Brenes et al., 2017; Johansen et al., 2018; Ok and Ozdarici-Ok, 2018). Dense 3D point clouds can be utilized for geometric measurements since they are adequately representative (de Castro et al., 2019). De Castro et al. (2018) also observed similar differences in the quality of 3D reconstructions of tree canopies produced by two training methods: hedgerow and intense. De Castro et al. (2018) studied a novel object-based image analysis approach based on DSM for 3D characterization of grapevines in the context of precision viticulture in order to establish site-specific management strategies. Mammarella et al. (2022) used a novel approach in which a fleet of UAVs performed remote sensing tasks over an apple orchard to reconstruct a 3D map of the field. In doing so, they formulated the coverage control problem to combine the position of a surveillance target and the viewing angle. The improvement of autonomous driving capabilities have been also investigated by automating the process of retrieving low–complexity maps from data collected with preliminary remote sensing missions, and making them available for autonomous navigation within orchards by unmanned ground vehicles (Mammarella et al., 2021; Katikaridis et al., 2022).

Pest management and drone spraying

Diseases can occur at any time of the fruit growing season, from flowering to harvest and even during the winter dormancy of the trees. Drones have already been used in orchards to monitor a wide range of diseases, but their suitability for monitoring severe diseases needs further research. However, identifying diseases using aerial drone imagery allows orchards to be inspected over a wider area, while saving money on labor and equipment (Radoglou-Grammatikis et al., 2020).

The first step in disease control is to identify the disease. The difficulty of disease diagnosis and the diversity of fruit species make it difficult to apply research results to other fruit tree disease detection technologies. Different sensors have been used to classify disease signs so that valuable indicators can be derived from aerial photographs. The potential of UAVs equipped with hyperspectral and multispectral sensors to categorize citrus trees infected with two types of biotic diseases, bacterial canker and Huang-Longbing (HLB), was described by Garcia-Ruiz et al. (2013) and Abdulridha et al. (2019). Sankaran et al. (2013) developed and tested a drone application to evaluate the ability of two cameras to monitor the health of citrus orchards. The field of view, classification accuracy, and image resolution of the two cameras were evaluated. Experiments were conducted in Florida citrus orchards where some trees were healthy and others had citrus greening or HLB.

Apple scab has become a major problem in apple orchards, affecting fruit yield and quality. The use of UAV technology to monitor apple scab in orchards has the potential to provide a model for risk assessment (Stella et al., 2017). Leaf moisture data collection can be used to indirectly monitor apple scab. Data collected by drones on leaf wetness data were used to provide accurate inputs to the risk prediction assessment model. Detection technologies based on thermal or spectral sensors are displacing traditional detection methods that rely on assessment by the farmer's own eye. In addition, RGB sensors have the potential to be used to detect diseases with visual symptoms.

Manual spraying exposes personnel to a high-risk hazardous chemical environment. Drone-based systems for spraying pesticides in orchards are safer, more accurate, and less expensive than manual spraying or manned agricultural aircraft, according to the literature on precision agriculture (Martinez-Guanter et al., 2020; Zhang et al., 2017). UAV technology has proven useful in crop protection and pest control not only for observation and detection, but also for precise application of agrochemical sprays, where the applicator can be moved away from the spray house and spraying can be done with highly targeted spatial resolution, especially in difficult geographic terrains (Giles and Billing, 2015). Precision aerial applications on small targets are possible with the use of small UAVs equipped with spray systems. Recently, drones have been used for variable rate application, including selective treatment of ultra-low volume pesticides (Martinez-Guanter et al., 2020). These precision applications have the potential to save money and reduce danger to operators during treatments.

The spraying application on the so-called "specialty crops" (such as vineyards, orchards, citrus, and olive trees) is one of the most controversial and important activities with direct economic, technical, and environmental impacts. Campos et al. (2019) used a drone equipped with a multispectral camera to obtain data for a map of canopy growth of an entire plot. The canopy map was then converted into a practical prescription map, which was then entered into the specific software of the sprayer. Compared to conventional spraying application, site-specific management for spray treatment in vineyards resulted in a 45% reduction in application rates, based on this study. Martinez-Guanter et al. (2020) designed and developed a small application system that can be mounted on a drone for agrochemical spraying applications, as well as analyzed the application quality and economic cost in olive and citrus orchards compared to those of conventional treatment. According to the results, the application costs for the aerial vehicle and conventional equipment differed by about seven euros per hectare. Finally, Sarri et al. (2019) studied the spraying performance of a commercial drone equipped with different types of nozzles and compared it with sprayers commonly used in small vineyards in the mountains. Field trials were conducted in a small terraced vineyard on a high hill. Data on droplet coverage, density, and size were collected and analyzed using water-sensitive papers attached to the leaves with clips. According to the results, the operating capacity of the drone was twice that of the spray gun and 1.6 times that of the bag sprayer. The position of the targets (water-sensitive leaves) and the type of sprayer used had an effect on droplet coverage, density, and size.

However, aerial spraying can be useless without a realistic spraying plan and a thorough and accurate information support, such as identifying parts of the tree canopy that have been characterized as target areas for spraying. Drone spraying, unlike other drone-based management tasks, has fewer operational parameters, such as spray nozzle control, altitude, and speed. A change in wind direction or deviation from the flight path has a significant impact on the uniformity of droplet distribution. Even when using the same spray technology, tree shape should be considered as an operational factor (Zhang et al., 2016).

The parameters related to the specific and particular characteristics of the tree canopy are the most important to control for better and more effective spraying (structure, dimensions, trellis system, etc.) (Balsari et al., 2008; Rosell and Sanz, 2012; Salcedo et al., 2015). Several studies have already been conducted with the aim of quantifying the relationship between spray quality and differences in canopy characteristics (Doruchowski et al., 2008; Gil et al., 2014a, 2014b; Miranda-Fuentes et al., 2016; Garcerá et al., 2017). In addition, several approaches have been presented in recent decades that not only serve to characterize tree canopy but also provide a good way to communicate the intended dose (Toews and Friessleben, 2012; Walklate and Cross, 2012).

Irrigation efficiency

An efficient irrigation strategy is key for minimizing the negative impact on fruit growers' profit. Irrigation scheduling is critical in orchards and vineyards because it directly affects yield and fruit quality, both of which depend on proper water use. Even drought-tolerant species such as olive can benefit from irrigation (*Olea europaea* L.), which can increase growth, production, and fruit quality (olive oil) when grown in high-density cropping systems (Egea et al., 2017; Caruso et al., 2019). Water supply should match the actual needs of the trees when deficit irrigation is introduced. Site-specific water management is critical to address the variability of water demand in orchards.

Water stress in plants is one of the most important variables in abiotic stress (Gautam and Pagay, 2020). Therefore among the agricultural applications of remote thermal sensing with drones, detection of water stress from crop temperature data is a crucial one (Messina and Modica, 2020). Precision irrigation management depends on the precision and practicality of sensor data used to analyze crop water status. Data from a multispectral and thermal infrared image acquired by an UAV were examined to determine if it could be used to evaluate various subsurface irrigation practices. To investigate potential indicators of water stress detection, various UAV-generated reflectance indices were calculated, evaluated, and compared with ground-based water stress testing methods. Various vegetation indexes, including NDVI, produced by integrating data from many spectral ranges have been shown to be potentially useful for predicting water status in orchards (Caruso et al., 2019). Water status has been shown to influence chlorophyll and fluorescence indices (at leaf level), green fraction (GR), extended normalized difference vegetation index (ENDVI), normalized near-infrared difference green index (NDGNI), and saturation (S) (Zarco-Tejada et al., 2013; Bulanon et al., 2016). Leaf area index (LAI), on the other hand, is a measure of canopy structure that is susceptible to water stress. Normalized vegetation index values estimated from UAV images are correlated with LAI measured on the ground, as shown by studies using LAI indirect measurement (Berni et al., 2009b; Caruso et al., 2019).

Thermal cameras carried by UAVs have been used in several studies to calculate crop water stress index (CWSI) in perennial plants. Bellvert et al. (2016) used a thermal camera mounted on a drone to map internal orchard geographic variability and to investigate the relationship between CWSI and leaf water potential over multiple growing seasons. Gonzalez-Dugo et al. (2013) studied the water status of five different fruit trees, namely: almond (*Prunus dulcis*), apricot (*Prunus armeniaca*), peach (*Prunus persica*), lemon (Citrus × limon), and orange (*Citrus sinensis*). In other research efforts, UAV-based thermal remote sensing was used to detect crown temperature for water stress in citrus orchards (oranges—*Citrus sinensis*, and mandarins—*Citrus reticulata*) (Zarco-Tejada et al., 2012; Gonzalez-Dugo et al., 2014). Gómez-Candón et al. (2016) presented a method for creating thermal orthomosaics and a method for radiometric correction of UAV-produced thermal data based on tests in an apple orchard.

Olive crops, a very important species in the Mediterranean region, have also been studied to evaluate plant behavior in response to different irrigation treatments (Solano et al., 2019). By establishing reliable links between the corresponding index and water stress indicators, such as stomatal conductance, stem water potential, and leaf transpiration rate, Egea et al. (2017) demonstrated the use of CWSI to monitor water stress in a dense olive orchard. Ortega-Farías et al. (2016) evaluated the components of energy balance in an olive grove with drip irrigation using thermal imaging and multispectral cameras mounted on drones capturing high-resolution images to quantify spatial variability within the field. Berni et al. (2009a) used high-resolution thermal data to calculate and map canopy conductance and CWSI in a heterogeneous olive orchard.

Finally, in vineyards, CWSI is also commonly used. The relationships between temperature-related indices obtained from thermal and multispectral images, the stomatal conductivity, and the water potential were established by Baluja et al. (2012). Santesteban et al. (2017) proposed a UAV application to analyze thermal images to predict seasonal and immediate variations in water balance in a vineyard. To measure the changes in water balance, CWSI values from UAV images were related to stem water potential and seasonal stomatal conductance of leaves. Espinoza et al. (2017) showed how multispectral and thermal imaging data from low altitude can be used to evaluate irrigation techniques and water stress in plants. Poblete-Echeverría et al. (2014) also investigated how well aircraft and ground-based infrared thermography can detect water stress in vineyards and olive groves. An UAV was used to capture infrared thermal images from a nadir position, while a handheld infrared camera was used to capture lateral infrared thermal images. Matese et al. (2018) and Pádua et al. (2019, 2020) used a variety of sensors (RGB, multispectral, and thermal) for a number of precision viticulture applications, including vigor mapping, multitemporal vigor mapping, and water stress detection.

Yield prediction and selective harvesting

One of the most critical issues in orchards is the nondestructive evaluation of fruit maturity, since many management decisions are closely related to fruit maturity status. Establishing a relationship between fruit maturity and visual appearance, such as color and texture, allows the development of new irrigation, harvesting, and postharvest techniques, as well as new yield forecasting methods based on growth analysis (Li et al., 2018; Sabzi et al., 2019).

Several studies have shown that UAV-based photography can help in data-driven methods for yield estimation (Liu et al., 2021). The most direct and accurate method for counting flowers/fruits on trees is using UAV imaging technology (Horton et al., 2017). Apolo-Apolo et al. (2020) presented a cloud-based environment for generating yield estimation maps from apple orchards using UAV images.

It has been suggested that optimizing harvest timing, fruit tree vigor, and health monitoring should be prioritized to maximize fruit yield and harvest operations (Vanbrabant et al., 2019). Fruit maturity is a critical factor in determining harvest timing, as it directly affects fruit quality during transport to markets. Initial tests of ethylene detection using ethylene-sensitive sensors attached to drones have identified the effects of flight altitude and wind speed measurement in assessing apple ripening (Valente et al., 2019).

Various authors have already used different imaging technologies in the field of nondestructive maturity analysis, including hyperspectral, multispectral, and visible light (Lu and Lu, 2017); (Arendse et al., 2018), the latter being the most favorable. However, the images are usually of high resolution, and there are few studies on predicting fruit ripeness from low-resolution photos taken by satellites or drones. Estimating ripeness in orchards helps improve postharvest activities. Harvesting fruits based on their maturity stage can reduce storage costs and improve market outcomes. Aerial imagery and estimated ripeness can also be used to guide water stress monitoring and irrigation water use evaluation (Sabzi et al., 2019).

Flight planning

Different combinations of flight factors affect the quality and usability of the final products, so determining the right setting for flight planning is critical (Dandois et al., 2015; Aasen et al., 2018; Roth et al., 2018; Dolias et al., 2022). Multispectral drone imagery has proven to be a beneficial tool for determining canopy structure and the parameters for the appropriate application. Optimizing flight planning settings is critical in this context, as they greatly affect the quality of the images and maps of tree and plant biophysical parameters produced, determining the optimal settings for flight planning is critical in this context. Variables such as flight altitude, image overlap, flight direction, flight speed, and sun position must be carefully considered during flight planning to produce the best drone images. The effects of individual variables on image quality have been studied in the past, but the interaction of numerous variables still needs to be explored. Tu et al. (2020) investigated the effects of different flight characteristics on data quality metrics at each processing stage, including photo alignment, point cloud summarization, 3D model construction, and orthomosaicking. The result of the research is a workflow for drone flight planning and image processing that enables accurate measurements of tree stands. This includes the quality of tie points, density of the compacted point cloud, and accuracy of height and projective plant cover measurements derived from individual trees in a commercial avocado plantation. Data quality was improved by flying along the hedgerow, at high sun elevation, and with low image tilt angles. To achieve the required forward overlap, the optimal flight speed must be determined. The effects of each variable for image acquisition are discussed in detail, and techniques for optimizing flight planning for three situations with different drone settings are proposed. Developing techniques for acquiring drone data over horticultural tree crops that provide excellent image capture would increase confidence in the accuracy of the following algorithms and biophysical property maps.

Conclusions

The objective of this chapter was to provide a concise and comprehensive overview of how drones are already being used in orchards and vineyards, as well as information on specific applications. The research presented varies in terms of objectives and equipment used, but they all show how different technologies can be used in combination with drones to determine the diversity of orchards and vineyards by analyzing structural aspects, disease incidence, and plant physiology. Drones thus appear to be a technology that can be used to improve decision-making in orchard and vineyard management. Improved disease diagnosis, and prevention, as well as prescription mapping and variable rate application of pesticides, will lead to less product waste and more sustainability.

UAVs have the ability to deliver exceptionally high spatial resolution at the expense of temporal resolution. They can fly over rough terrain, but many platforms, especially those with heavy payloads, have limited flight time. Weather conditions can also have a significant impact on UAV effectiveness. Thanks to the ability to fly different routes and incorporate depth sensors into UAVs, accurate 3D models and maps can be created. However, it should be emphasized that when using high-resolution images, photogrammetric processing can be complex and time consuming. UAVs can also physically interact with the environment, as for example, when collecting samples and applying agro-chemicals. UAVs can also be connected to an existing sensor network to provide more integrated precision agriculture plans. However, much of this sensor and software technology requires a significant upfront investment. Therefore the cost-benefit analysis of using UAVs must be carefully weighed against other options.

Although high efficiency in orchard and vineyard management can be beneficial to farmers and companies in the fruit industry, drones are underutilized in this sector overall. Despite the effective use of agricultural drones in some countries, application costs remain a major barrier to their widespread use in precision management of orchards and vineyards.

References

Aasen, H., Honkavaara, E., Lucieer, A., Zarco-Tejada, P.J., 2018. Quantitative remote sensing at ultra-high resolution with UAV spectroscopy: a review of sensor technology, measurement procedures, and data correction workflows. Remote Sens. 10 (7), 1091.

Abdulridha, J., Batuman, O., Ampatzidis, Y., 2019. UAV-based remote sensing technique to detect citrus canker disease utilizing hyperspectral imaging and machine learning. Remote Sens. 11 (11), 1373.

Apolo-Apolo, O.E., Pérez-Ruiz, M., Guanter, J.M., Valente, J., 2020. A cloud-based environment for generating yield estimation maps from apple orchards using UAV imagery and a deep learning technique. Front. Plant Sci. 1086.

Arendse, E., Fawole, O.A., Magwaza, L.S., Opara, U.L., 2018. Non-destructive prediction of internal and external quality attributes of fruit with thick rind: a review. J. Food Eng. 217, 11−23.

Balsari, P., Doruchowski, G., Marucco, P., Tamagnone, M., Van de Zande, J., Wenneker, M., 2008. A system for adjusting the spray application to the target characteristics. Agric. Eng. Int. 10.

Baluja, J., Diago, M.P., Balda, P., Zorer, R., Meggio, F., Morales, F., et al., 2012. Assessment of vineyard water status variability by thermal and multispectral imagery using an unmanned aerial vehicle (UAV). Irrig. Sci. 30 (6), 511−522.

Bellvert, J., Marsal, J., Girona, J., Gonzalez-Dugo, V., Fereres, E., Ustin, S.L., et al., 2016. Airborne thermal imagery to detect the seasonal evolution of crop water status in peach, nectarine and Saturn peach orchards. Remote Sens. 8 (1), 39.

Berni, J.A.J., Zarco-Tejada, P.J., Sepulcre-Cantó, G., Fereres, E., Villalobos, F., 2009a. Mapping canopy conductance and CWSI in olive orchards using high resolution thermal remote sensing imagery. Remote Sens. Environ. 113 (11), 2380−2388.

Berni, J.A., Zarco-Tejada, P.J., Suárez, L., Fereres, E., 2009b. Thermal and narrowband multispectral remote sensing for vegetation monitoring from an unmanned aerial vehicle. IEEE Trans. Geosci. Remote Sens. 47 (3), 722–738.

Bulanon, D.M., Lonai, J., Skovgard, H., Fallahi, E., 2016. Evaluation of different irrigation methods for an apple orchard using an aerial imaging system. ISPRS Int. J. Geoinform. 5 (6), 79.

Calderón, R., Navas-Cortés, J.A., Lucena, C., Zarco-Tejada, P.J., 2013. High-resolution airborne hyperspectral and thermal imagery for early detection of Verticillium wilt of olive using fluorescence, temperature and narrow-band spectral indices. Remote Sens. Environ. 139, 231–245.

Campos, J., Llop, J., Gallart, M., García-Ruiz, F., Gras, A., Salcedo, R., et al., 2019. Development of canopy vigour maps using UAV for site-specific management during vineyard spraying process. Precis. Agric. 20 (6), 1136–1156.

Caruso, G., Zarco-Tejada, P.J., González-Dugo, V., Moriondo, M., Tozzini, L., Palai, G., et al., 2019. High-resolution imagery acquired from an unmanned platform to estimate biophysical and geometrical parameters of olive trees under different irrigation regimes. PLoS One 14 (1), e0210804.

Castillo-Ruiz, F.J., Jimenez-Jimenez, F., Blanco-Roldán, G.L., Sola-Guirado, R.R., Agueera-Vega, J., Castro-Garcia, S., 2015. Analysis of fruit and oil quantity and quality distribution in high-density olive trees in order to improve the mechanical harvesting process. Span. J. Agric. Res. 13 (2), e0209.

De Castro, A.I., Jiménez-Brenes, F.M., Torres-Sánchez, J., Peña, J.M., Borra-Serrano, I., López-Granados, F., 2018. 3-D characterization of vineyards using a novel UAV imagery-based OBIA procedure for precision viticulture applications. Remote Sen. 10 (4), 584. Available from: https://doi.org/10.3390/rs10040584.

de Castro, A.I., Rallo, P., Suárez, M.P., Torres-Sánchez, J., Casanova, L., Jiménez-Brenes, F.M., et al., 2019. High-throughput system for the early quantification of major architectural traits in olive breeding trials using UAV images and OBIA techniques. Front. Plant Sci. 1472.

Dandois, J.P., Olano, M., Ellis, E.C., 2015. Optimal altitude, overlap, and weather conditions for computer vision UAV estimates of forest structure. Remote Sens. 7 (10), 13895–13920.

Dolias, G., Benos, L., Bochtis, D., 2022. On the routing of unmanned aerial vehicles (UAVs) in precision farming sampling missions. In: Bochtis, D.D., Sørensen, C.G., Fountas, S., Moysiadis, V., Pardalos, P.M. (Eds.), Information and Communication Technologies for Agriculture—Theme III: Decision (Springer Optimization and Its Applications), 184. Springer.

Doruchowski, G., Balsari, P., Van De Zande, J., 2008. Development of a crop adapted spray application system for sustainable plant protection in fruit growing. Int. Symp. Appl. Precis. Agric. Fruits Veg. 824, 251–260.

Egea, G., Padilla-Díaz, C.M., Martinez-Guanter, J., Fernández, J.E., Pérez-Ruiz, M., 2017. Assessing a crop water stress index derived from aerial thermal imaging and infrared thermometry in super-high density olive orchards. Agric. Water Manag. 187, 210–221.

Espinoza, C.Z., Khot, L.R., Sankaran, S., Jacoby, P.W., 2017. High resolution multispectral and thermal remote sensing-based water stress assessment in subsurface irrigated grapevines. Remote Sens. 9 (9), 961.

Gago, J., Douthe, C., Coopman, R.E., Gallego, P.P., Ribas-Carbo, M., Flexas, J., et al., 2015. UAVs challenge to assess water stress for sustainable agriculture. Agric. Water Manag. 153, 9–19.

Garcerá, C., Fonte, A., Moltó, E., Chueca, P., 2017. Sustainable use of pesticide applications in citrus: a support tool for volume rate adjustment. Int. J. Environ. Res. Public Health 14 (7), 715.

Garcia-Ruiz, F., Sankaran, S., Maja, J.M., Lee, W.S., Rasmussen, J., Ehsani, R., 2013. Comparison of two aerial imaging platforms for identification of Huanglongbing-infected citrus trees. Comput. Electron. Agric. 91, 106−115.

Gautam, D., Pagay, V., 2020. A review of current and potential applications of remote sensing to study the water status of horticultural crops. Agronomy 10 (1), 140.

Gil Moya, E., Gallart González-Palacio, M., Llorens Calveras, J., Llop Casamada, J., Bayer, T., Carvalho, C., 2014b. Spray adjustments based on LWA concept in vineyard. Relationship between canopy and coverage for different application settings. Asp. Appl. Biol. Int. Adv. Pesticide Appl. 122, 25−32.

Giles, D., Billing, R., 2015. Deployment and performance of a UAV for crop spraying. Chem. Eng. Trans. 44, 307−312.

Gil, E., Arnó, J., Llorens, J., Sanz, R., Llop, J., Rosell-Polo, J.R., et al., 2014a. Advanced technologies for the improvement of spray application techniques in Spanish viticulture: an overview. Sensors 14 (1), 691−708.

Gonzalez-Dugo, V., Zarco-Tejada, P.J., Fereres, E., 2014. Applicability and limitations of using the crop water stress index as an indicator of water deficits in citrus orchards. Agric. For. Meteorol. 198, 94−104.

Gonzalez-Dugo, V., Zarco-Tejada, P., Nicolás, E., Nortes, P.A., Alarcón, J.J., Intrigliolo, D.S., et al., 2013. Using high resolution UAV thermal imagery to assess the variability in the water status of five fruit tree species within a commercial orchard. Precis. Agric. 14 (6), 660−678.

Gómez-Candón, D., Virlet, N., Labbé, S., Jolivot, A., Regnard, J.L., 2016. Field phenotyping of water stress at tree scale by UAV-sensed imagery: new insights for thermal acquisition and calibration. Precis. Agric. 17 (6), 786−800.

Hassler, S.C., Baysal-Gurel, F., 2019. Unmanned aircraft system (UAS) technology and applications in agriculture. Agronomy 9 (10), 618.

Horton, R., Cano, E., Bulanon, D., Fallahi, E., 2017. Peach flower monitoring using aerial multispectral imaging. J. Imag. 3 (1), 2.

Jiménez-Brenes, F.M., López-Granados, F., De Castro, A.I., Torres-Sánchez, J., Serrano, N., Peña, J.M., 2017. Quantifying pruning impacts on olive tree architecture and annual canopy growth by using UAV-based 3D modelling. Plant Methods 13 (1), 1−15.

Johansen, K., Raharjo, T., McCabe, M.F., 2018. Using multi-spectral UAV imagery to extract tree crop structural properties and assess pruning effects. Remote Sens. 10 (6), 854.

Kasimati, A., Espejo-Garcia, B., Vali, E., Malounas, I., Fountas, S., 2021a. Investigating a selection of methods for the prediction of total soluble solids among wine grape quality characteristics using normalized difference vegetation index data from proximal and remote sensing. Front. Plant Sci. 12, 1118.

Kasimati, A., Kalogrias, A., Psiroukis, V., Grivakis, K., Taylor, J.A., Fountas, S., 2021b. Are all NDVI maps created equal−comparing vineyard NDVI data from proximal and remote sensing. Precis. Agric. 21, 1366−1376.

Katikaridis, D., Moysiadis, V., Tsolakis, N., Busato, P., Kateris, D., Pearson, S., et al., 2022. UAV-supported route planning for UGVs in semi-deterministic agricultural environments. Agronomy 12 (8), 1937.

Liu, J., Xiang, J., Jin, Y., Liu, R., Yan, J., Wang, L., 2021. Boost precision agriculture with unmanned aerial vehicle remote sensing and edge intelligence: a survey. Remote Sens. 13 (21), 4387.

Li, B., Lecourt, J., Bishop, G., 2018. Advances in non-destructive early assessment of fruit ripeness towards defining optimal time of harvest and yield prediction—a review. Plants 7 (1), 3.

Lu, Y., Lu, R., 2017. Non-destructive defect detection of apples by spectroscopic and imaging technologies: a review. Trans. ASABE 60 (5), 1765−1790.

Ludovisi, R., Tauro, F., Salvati, R., Khoury, S., Mugnozza Scarascia, G., Harfouche, A., 2017. UAV-based thermal imaging for high-throughput field phenotyping of black poplar response to drought. Front. Plant Sci. 8, 1681.

Mammarella, M., Comba, L., Biglia, A., Dabbene, F., Gay, P., 2021. Cooperation of unmanned systems for agricultural applications: a theoretical framework. Biosyst. Eng. 223.

Mammarella, M., Donati, C., Shimizu, T., Suenaga, M., Comba, L., Biglia, A. et al., 2022. 3D map reconstruction of an orchard using an angle-aware covering control strategy. arXiv preprint arXiv:2202.02758.

Martinez-Guanter, J., Agüera, P., Agüera, J., Pérez-Ruiz, M., 2020. Spray and economics assessment of a UAV-based ultra-low-volume application in olive and citrus orchards. Precis. Agric. 21 (1), 226–243. Available from: https://doi.org/10.1007/s11119-019-09665-7.

Matese, A., Di Gennaro, S.F., Berton, A., 2017. Assessment of a canopy height model (CHM) in a vineyard using UAV-based multispectral imaging. Int. J. Remote Sens. 38 (8–10), 2150–2160.

Matese, A., Baraldi, R., Berton, A., Cesaraccio, C., Di Gennaro, S.F., Duce, P., et al., 2018. Estimation of water stress in grapevines using proximal and remote sensing methods. Remote Sens. 10 (1), 114.

Messina, G., Modica, G., 2020. Applications of UAV thermal imagery in precision agriculture: state of the art and future research outlook. Remote Sens. 12 (9), 1491.

Miranda-Fuentes, A., Llorens, J., Gamarra-Diezma, J.L., Gil-Ribes, J.A., Gil, E., 2015. Towards an optimized method of olive tree crown volume measurement. Sensors 15 (2), 3671–3687.

Miranda-Fuentes, A., Llorens, J., Rodríguez-Lizana, A., Cuenca, A., Gil, E., Blanco-Roldán, G.L., et al., 2016. Assessing the optimal liquid volume to be sprayed on isolated olive trees according to their canopy volumes. Sci. Total Environ. 568, 296–305.

Murugan, D., Garg, A., Singh, D., 2017. Development of an adaptive approach for precision agriculture monitoring with drone and satellite data. IEEE J. Sel. Top. Appl. Earth Observ. Remote Sens. 10 (12), 5322–5328.

Navia, J., Mondragon, I., Patino, D., Colorado, J., 2016. Multispectral mapping in agriculture: terrain mosaic using an autonomous quadcopter UAV. In: 2016 International Conference on Unmanned Aircraft Systems (ICUAS), 1351–1358. IEEE.

Nolan, A.P., Park, S., O'Connell, M., Fuentes, S., Ryu, D., Chung, H., 2015. Automated detection and segmentation of vine rows using high resolution UAS imagery in a commercial vineyard. In: International Congress on Modelling and Simulation 2015: Partnering with industry and the community for innovation and impact through modelling, 1406–1412. Modelling and Simulation Society of Australia and New Zealand Inc. (MSSANZ).

Ok, A.O., Ozdarici-Ok, A., 2018. Combining orientation symmetry and LM cues for the detection of citrus trees in orchards from a digital surface model. IEEE Geosci. Remote Sens. Lett. 15 (12), 1817–1821.

Ortega-Farías, S., Ortega-Salazar, S., Poblete, T., Kilic, A., Allen, R., Poblete-Echeverría, C., et al., 2016. Estimation of energy balance components over a drip-irrigated olive orchard using thermal and multispectral cameras placed on a helicopter-based unmanned aerial vehicle (UAV). Remote Sens. 8 (8), 638.

Poblete-Echeverría, C., Sepulveda-Reyes, D., Ortega-Farias, S., Zuñiga, M., Fuentes, S., 2014. Plant water stress detection based on aerial and terrestrial infrared thermography: a study case from vineyard and olive orchard. In: XXIX International Horticultural Congress on Horticulture: Sustaining Lives, Livelihoods and Landscapes (IHC2014): 1112, 141–146.

Pádua, L., Marques, P., Adão, T., Guimarães, N., Sousa, A., Peres, E., et al., 2019. Vineyard variability analysis through UAV-based vigour maps to assess climate change impacts. Agronomy 9 (10), 581.

Pádua, L., Adão, T., Sousa, A., Peres, E., Sousa, J.J., 2020. Individual grapevine analysis in a multi-temporal context using UAV-based multi-sensor imagery. Remote Sens. 12 (1), 139.

Radoglou-Grammatikis, P., Sarigiannidis, P., Lagkas, T., Moscholios, I., 2020. A compilation of UAV applications for precision agriculture. Comput. Netw. 172, 107148.

Rom, C.R., 1991. Light thresholds for apple tree canopy growth and development. HortScience 26 (8), 989−992.

Rosell, J.R., Sanz, R., 2012. A review of methods and applications of the geometric characterization of tree crops in agricultural activities. Comput. Electron. Agric. 81, 124−141.

Roth, L., Streit, B., 2017. Predicting cover crop biomass by lightweight UAS-based RGB and NIR photography: an applied photogrammetric approach. Precis. Agric. 19 (1), 93−114.

Roth, L., Hund, A., Aasen, H., 2018. PhenoFly planning tool: flight planning for high-resolution optical remote sensing with unmanned areal systems. Plant Methods 14 (1), 1−21.

Sabzi, S., Abbaspour-Gilandeh, Y., García-Mateos, G., Ruiz-Canales, A., Molina-Martínez, J.M., Arribas, J.I., 2019. An automatic non-destructive method for the classification of the ripeness stage of red delicious apples in orchards using aerial video. Agronomy 9 (2), 84.

Salcedo, R., Garcera, C., Granell, R., Molto, E., Chueca, P., 2015. Description of the airflow produced by an air-assisted sprayer during pesticide applications to citrus. Span. J. Agric. Res. 13 (2), e0208.

Sankaran, S., Khot, L.R., Maja, J.M., Ehsani, R., 2013. Comparison of two multiband cameras for use on small UAVs in agriculture. In: 2013 5th Workshop on Hyperspectral Image and Signal Processing: Evolution in Remote Sensing (WHISPERS), 1−4. IEEE.

Santesteban, L.G., Di Gennaro, S.F., Herrero-Langreo, A., Miranda, C., Royo, J.B., Matese, A., 2017. High-resolution UAV-based thermal imaging to estimate the instantaneous and seasonal variability of plant water status within a vineyard. Agric. Water Manag. 183, 49−59.

Sarri, D., Martelloni, L., Rimediotti, M., Lisci, R., Lombardo, S., Vieri, M., 2019. Testing a multi-rotor unmanned aerial vehicle for spray application in high slope terraced vineyard. J. Agric. Eng. 50 (1), 38−47.

Solano, F., Di Fazio, S., Modica, G., 2019. A methodology based on GEOBIA and WorldView-3 imagery to derive vegetation indices at tree crown detail in olive orchards. Int. J. Appl. Earth Observ. Geoinf. 83, 101912.

Stella, A., Caliendo, G., Melgani, F., Goller, R., Barazzuol, M., La Porta, N., 2017. Leaf wetness evaluation using artificial neural network for improving apple scab fight. Environments 4 (2), 42.

Surový, P., Almeida Ribeiro, N., Panagiotidis, D., 2018. Estimation of positions and heights from UAV-sensed imagery in tree plantations in agrosilvopastoral systems. Int. J. Remote Sens. 39 (14), 4786−4800.

Tagarakis, A.C., Filippou, E., Kalaitzidis, D., Benos, L., Busato, P., Bochtis, D., 2022. Proposing UGV and UAV systems for 3D mapping of orchard environments. Sensors 22, 1571.

Toews, R.B., Friessleben, R., 2012. Dose rate expression—need for harmonization and consequences of the leaf wall area approach. Erwerbs-Obstbau 54 (2), 49−53.

Torres-Sánchez, J., López-Granados, F., Serrano, N., Arquero, O., Peña, J.M., 2015. High-throughput 3-D monitoring of agricultural-tree plantations with unmanned aerial vehicle (UAV) technology. PLoS One 10 (6), e0130479.

Torres-Sanchez, J., de Castro, A.I., Pena, J.M., Jimenez-Brenes, F.M., Arquero, O., Lovera, M., et al., 2018a. Mapping the 3D structure of almond trees using UAV acquired photogrammetric point clouds and object-based image analysis. Biosyst. Eng. 176, 172–184.

Torres-Sánchez, J., López-Granados, F., Borra-Serrano, I., Peña, J.M., 2018b. Assessing UAV-collected image overlap influence on computation time and digital surface model accuracy in olive orchards. Precis. Agric. 19 (1), 115–133.

Tu, Y.H., Phinn, S., Johansen, K., Robson, A., Wu, D., 2020. Optimising drone flight planning for measuring horticultural tree crop structure. ISPRS J. Photogramm. Remote Sens. 160, 83–96.

Valente, J., Almeida, R., Kooistra, L., 2019. A comprehensive study of the potential application of flying ethylene-sensitive sensors for ripeness detection in apple orchards. Sensors 19 (2), 372.

Vanbrabant, Y., Tits, L., Delalieux, S., Pauly, K., Verjans, W., Somers, B., 2019. Multitemporal chlorophyll mapping in pome fruit orchards from remotely piloted aircraft systems. Remote Sens. 11 (12), 1468.

Walklate, P.J., Cross, J.V., 2012. An examination of leaf-wall-area dose expression. Crop. Prot. 35, 132–134.

Yuan, W., Choi, D., 2021. UAV-based heating requirement determination for frost management in apple orchard. Remote Sens. 13 (2), 273.

Zarco-Tejada, P.J., González-Dugo, V., Berni, J.A., 2012. Fluorescence, temperature and narrow-band indices acquired from a UAV platform for water stress detection using a micro-hyperspectral imager and a thermal camera. Remote Sens. Environ. 117, 322–337.

Zarco-Tejada, P.J., Morales, A., Testi, L., Villalobos, F.J., 2013. Spatio-temporal patterns of chlorophyll fluorescence and physiological and structural indices acquired from hyperspectral imagery as compared with carbon fluxes measured with eddy covariance. Remote Sens. Environ. 133, 102–115.

Zarco-Tejada, P.J., Diaz-Varela, R., Angileri, V., Loudjani, P., 2014. Tree height quantification using very high resolution imagery acquired from an unmanned aerial vehicle (UAV) and automatic 3D photo-reconstruction methods. Eur. J. Agron. 55, 89–99.

Zhang, P., Deng, L., Lyu, Q., He, S.L., Yi, S.L., Liu, Y., et al., 2016. Effects of citrus tree-shape and spraying height of small unmanned aerial vehicle on droplet distribution. Int. J. Agric. Biol. Eng. 9, 45–52. Available from: https://doi.org/10.3965/j.ijabe.20160904.2178.

Zhang, P., Wang, K.J., Lyu, Q., He, S.L., Yi, S.L., Xie, R.J., et al., 2017. Droplet distribution and control against citrus leafminer with UAV spraying. Int. J. Robot. Autom. 32 (3), 299–307.

Zhang, C., Valente, J., Kooistra, L., Guo, L., Wang, W., 2019. Opportunities of UAVS in orchard management. Int. Arch. Photogramm. Remote Sens. Spatial Inform. Sci.

Zhang, C., Valente, J., Kooistra, L., Guo, L., Wang, W., 2021. Orchard management with small unmanned aerial vehicles: a survey of sensing and analysis approaches. Precis. Agric. 22 (6), 2007–2052.

SECTION C

Operational aspects

CHAPTER 6

Unmanned aerial vehicles for agricultural automation

Georgios Siavalas, Eleni Vrochidou and Vassilis G. Kaburlasos
Human-Machines Interaction (HUMAIN) Lab, Department of Computer Science, International Hellenic University (IHU), Kavala, Greece

Introduction

Precision agriculture (PA) refers to the improvement of agricultural yield as well as the decision-supported farm management by innovative sensors and data analysis tools. PA is adopted in agricultural production to optimize both yield quality and yield quantity. PA relies on big data information from monitoring plant health, growth, in-field environmental conditions, irrigation, yield, etc., so as to support timely and effective interventions. Note that precise crop monitoring is time consuming as well as costly since it requires constant sampling on vast fields and consistent recording of multiple data (Cisternas et al., 2020). The latter can be addressed with remote sensing systems.

Remote sensing can address the above requirements, since it can provide with multiple information in a fast and nondestructive way. Remote sensing is considered a major supportive technology for PA. More specifically, numerous crop-related sensory data can be monitored and used to support sustainable farm management decisions (Maes and Steppe, 2019). Recently, unmanned aerial vehicles (UAVs) are used for remote sensing applications, as a more effective alternative to ground systems. To date, the most extensive use of UAVs has been in military applications, such as combat and surveillance, due to their ability to perform high level, high-risk tasks by remote control. In recent years, UAVs have attracted the attention of both the academic and the industrial community due to their wide range of applications and advantages: low costs, high flexibility, ease of operation, high spatial resolution, multisensory data acquisition. More specifically, in the agricultural sector, UAVs have introduced a useful alternative for extracting phenotypic information of crops rapidly as well as nondestructively (Tsouros et al., 2019a). Low altitude UAV flights,

Unmanned Aerial Systems in Agriculture
DOI: https://doi.org/10.1016/B978-0-323-91940-1.00006-2

combined with powerful sensors, can provide rich information, that is, high-resolution images of crops, which improve the performance of data processing algorithms. Many different sensors can be mounted on Uavs, depending on the application, to provide multiple inputs to the crop management system (Tsouros et al., 2019b).

The objective of this work is to (1) provide adequate knowledge regarding UAVs characteristics and main sensors; (2) present basic UAV control functions such as mission planning; (3) analyze mosaicing methods and feature extraction methods;and (4) provide an up-to-date overview of UAVs most common applications in agriculture, from 2018 onward. Moreover, (5) review both single UAV and multi-UAV applications, and (6) highlight limitations and future trends. This work can serve as an updated guide for researchers interested in UAVs applications in PA, since it is a research area that is rapidly advancing.

The rest of the paper is structured as follows: Section "Chronicle of unmanned aerial vehicles" presents the chronicle of UAV development. Section "Unmanned aerial vehicle types and characteristics" reviews UAV types and basic characteristics in terms of movements, components, and sensors. Section "Mission planning" analyzes mission planning processes. Section "Mosaicing methods" presents mosaicing methods. Section "Data processing" discusses data processing toward feature extraction from UAV-acquired data. Section "Unmanned aerial vehicles in precision agriculture" provides an overview of UAV applications in PA, from 2018 onward; both single UAV and multi-UAV systems are included. Section "Discussion" discusses the results of this review, and describes limitations and future trends regarding UAVs in PA. Finally, Section "Conclusion" concludes by summarizing the contribution of this work.

Chronicle of unmanned aerial vehicles

During World War I, in 1903, the need for UAV emerged toward carrying explosive warheads in order to attack enemy targets. Unmanned aircrafts would be a solution to avoid the loss of experienced pilots whose training was time consuming but their need during the war immediate (Clark, 2000). Thus remotely controlled aircrafts for carrying explosives and hitting targets as well as for battleship surveillance were developed. Autopilot control was developed by the American Elmer Sperry in 1917; the corresponding prototype aircraft made its first test flight with a pilot inside the cockpit, who was in charge only for takeoff and landing while the rest of the flight was handled by the autopilot (Palik and Nagy, 2019).

After World War I, during interwar years, and as military technology improved aircrafts, smaller, faster, less costly, and able to maneuver aircrafts were developed due to the need of the pilotless target aircraft (PTA) program for realistic gunnery practice. The PTA program has produced the Queen Bee (Fig. 6.1A) in 1935. Queen Bee was the first UAV capable of returning home after completing its mission based on radio-controlled technologies. Queen Bee was the first UAV also called "Drone," named after the male bee. In 1941 the first successful live attack with a remotely radio-controlled UAV took place.

In 1944, during World War II, after several failed attempts due to navigation inaccuracies and technical issues, the TDR-1s achieved a 50% successful hit rate, boosting UAVs into a new era. After World War II, the interest shifted mainly to the design of UAVs. In 1948 a jet-propelled UAV, the XQ-2, was launched, mainly for gunnery practice. The latter was the ancestor of the well-known Ryan Firebee (Fig. 6.1B) that dominated during the Cold War. Ryan Firebee was able to collect precise

Figure 6.1 Historical unmanned aerial vehicles: (A) Queen Bee (1935) (Wikipedia, 2023b), (B) Ryan Firebee (1951) (Wikipedia, 2023a), (C) Predator (1994) (Airforce Technology, 2022), (D) Prime Air (2016) (Kurogbangkaw, 2016).

intelligence out of images and used for remotely guided photographic surveillance missions. At that time, drones were upgraded from aerial targets in reconnaissance UAVs. In Vietnam War, the use of UAVs was crucial, leading to major modifications and improvements to their system architecture.

Wars funded research on UAVs, therefore their most extensive use has been for military purposes. However, ever since a high-tech device became available to the public, many potential applications became clear. In order to adapt military technology to civilian use, many significant changes took place, for example, innovations in the landing and takeoff systems, aircraft design, sensing, for example, cameras (Martinez Leon et al., 2021).

In the 1970s, Abraham Karem immigrated to the United States of America where he founded an institution namely "Leading Systems Inc." that manufactured drones. The design of the model "Amber," latter, evolved to the famous Predator Drone (Fig. 6.1C), giving Karem the name "Drone Father." Drone manufacturing industry largely advanced in 2005, by massive technological improvements that resulted in cheaper, smaller, and lightweight electronic sensors, for example, inertial measurement unit (IMU), global positioning systems (GPS), and cameras. Until that time, drones were restricted to military use (Békési et al., 2016). Only in 2005 the legislation on air traffic rules and the Federal Aviation Administration (FAA) officially issued the first commercial drone license. FAA regulations are complicated and keep changing; the development of regulations for nongovernmental drones into public airspace is an ongoing process.

In 2020 Amazon, having already its first package delivery with Prime Air (Fig. 6.1D), announced that drones were approved by FAA to deliver packages. In the meantime, many companies commercialize drones. For almost a decade, drone companies developed and updated their models, launching regularly models with extra features like quality cameras, powerful batteries, and IMUs. In recent years, drones are adapted to civil and academic usage, covering a wide range of applications from pleasure to research. Historical UAVs are illustrated in Fig. 6.1.

The growing interest in the use of drones was supported by technological advancements, by the wide range of potential applications, the relatively low-cost yet precise devices, and the prospect for significant economic impact in an ever-evolving field. Drone applications appear in robotic research in higher education, in crop management, in environmental monitoring, in firefighting, to cover sports and events, in aerial photography, in art, cinema, etc. (Morris, 2015). Therefore different drones were designed for different applications, for example, "Agras" from DJI has been used in PA, other drones were designed for transportation, mapping, cinematic videos, etc.

It should be noted that most commercial drones have one basic common characteristic: they are quadcopters. Note that quadcopters were introduced in 1908. Vertical takeoff and landing (VTOL) was designed and tested by Louis Breguet, who developed a four-rotor helicopter that was able to fly to an altitude of a few feet, yet, without providing concrete evidence or further details. At the end of 1923, Dr. George de Bothezat and Ivan Jerome developed the De Bothezat helicopter which was developed by the US Air Service. After around 100 flights, its highest reached altitude was 5 m above the ground (Mohammad Zain et al., 2002).

Unmanned aerial vehicle types and characteristics

This section presents types and characteristics of UAVs. First, basic UAV categories are defined based on their operational duration and radius. Next, characteristics of the movement of UAVs are analyzed. Then, basic components of a UAV system are presented, and finally, the most common UAV sensors in PA applications are reviewed.

Unmanned aerial vehicles categories

UAVs can be classified based on several parameters such as maximum takeoff weight, operation altitude, level of autonomy, launching method, ownership, and airspace class (Yan et al., 2019). According to operational duration and radius, the following main classes can be distinguished:

- Long-endurance and high-altitude UAVs. The latter are mainly used in reconnaissance, or for interceptions/attacks. Well-known drones of this category are the Northrop Grumman RQ-4 Global Hawk and the General Atomics MQ-1 Predator (Gupta et al., 2016).
- Medium-range UAVs. The latter have an operation radius of around 650 km and are mainly used for reconnaissance and assessment of combat effects (Bange et al., 2021).
- Short-range small UAVs. The latter have an operation radius of around 350 km and operation height of around 3 km. Their autonomy is around 8—12 hours. Known drone of this class are the BAE Systems British Phoenix (Guettier et al., 2009).

- Close-range UAVs. These are capable of flights with limited duration from 1 to 6 hours based on the mission, while their operation span is around 30 km (Yan et al., 2019).
- Close-range/low-cost UAVs. Their operation radius is limited to 5 km. Well-known commercial drone of this category is the AeroVironment RQ-14 Dragon Eye (Cai et al., 2014).
- Mini drones. These are commercial drones with limited operation range, mainly addressed to amateurs. Usually, they are controlled by either a remote controller or a smart application displayed on portable devices such as mobile phones or tablets (Yan et al., 2019).

Unmanned aerial vehicle movement characteristics

Multicopter or multirotor is a rotorcraft having more than two lift-generating rotors. The most common type of a multicopter is the quadcopter which includes four rotors, one on each arm. A multicopter is also referred to as X4 or X8. While X4 is the usual quadcopter, X8 shares the same configuration as a quadcopter but with eight rotors, two on each arm; one rotor is facing up and the other one is facing down. This configuration is referred to as coaxial rotor and it offers more power and stability. In addition to quadcopters, several models of multicopters exist, such as, tricopters, hexacopters, and octacopters (Mfiri et al., 2016).

One of the main characteristics of a quadcopter regards its movements. More specifically, a quadcopter has the following four basic movements (Khuwaja et al., 2018):

- Throttle: Throttle control regards the ability of a quadcopter to change its altitude or to simply maintain it. In case an increase in altitude is pursued, all motors must simultaneously increase their speed. Respectively, a decrease in altitude requires a simultaneous decrease of speed of all motors (Fig. 6.2A).
- Pitch: Pitch control regards the drone's ability to move forward or backward. On one hand, moving forward is pursued when the two rear motors increase their speed, whereas the two front ones maintain it; on the other hand, backward movement requires the opposite movements, as illustrated in Fig. 6.2B.
- Roll: Roll control regards the movement to the left or to the right. On one hand, left movement is effected when the speed of the two right motors increases, whereas the speed of the two left motors is maintained; on the other hand, movement to the right is effected by

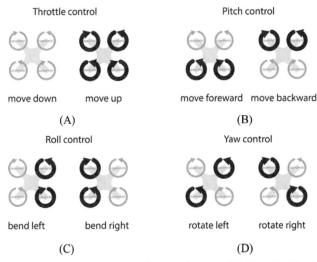

Figure 6.2 Basic movement controls of a quadcopter: (A) throttle, (B) pitch, (C) roll, and (D) yaw.

increasing the speed of the two left motors and maintaining the speed of the two right motors (Fig. 6.2C).

- Yaw: Yaw control regards the ability to rotate to the left or to the right. This can be achieved by increasing the speed of the diagonal motors front right and rear left and maintaining the speed of the other two motors for the left rotation. The opposite motor control is needed for right rotation (Fig. 6.2D).

From Fig. 6.2, it is obvious that all motors do not rotate in the same direction; more specifically, two of them are rotating clockwise (CW), whereas the other two are rotating counterclockwise (CCW). The latter practice is applied because each motor produces a rotational torque as it spins. If all motors were spinning CW, this would result in a yaw motion with a right rotation; whereas, spinning the motors in the opposite direction cancels the rotational torque of each motor, therefore the drone stays in place.

Unmanned aerial vehicle components

The basic components of a drone are the following (Magnussen et al., 2015; Parihar et al., 2016):

- Transmitter and receiver: The transceiver system is responsible for the communication between the controller and the drone. Through this system, all commands are sent from the operator to the drone, and the drone sends back live streaming or images to the controller, if a camera is onboard.

- Flight controller: The flight controller is the main component of the drone, just like a CPU is for a computer, or the brain for a human. It controls all functionalities of the drone such as motors speed, flight angle, and height.
- Electronic speed controller (ESC): ESC controls motor speed. Every ESC can handle only one motor. Therefore for a quadcopter (X4), four ESCs are needed.
- Brush-less direct current (DC) motor (BLDC): BLDC refers to the drone engines. A BLDC motor can achieve many revolutions per minute (RPM), for example, 30,000 RPM in order to provide the necessary thrust via the propellers to lift the craft. BLDC motors are closely related by kilovolts (kV). In general, the more the kV, the faster a motor can spin.
- Propellers: Propellers can be either pushers or pullers. For instance, a X4 drone has two CW spinning motors and two CCW spinning motors. A pusher propeller can produce thrust in CW rotation and a puller propeller in CCW rotation.
- Battery: Battery is a drone's power source. Depending on the size of the drone and the application, a battery can be either lithium polymer battery (LiPo) or lithium-ion battery (Li-ion).

Apart from the above basic components, a drone can be equipped with a variety of extra features and modules depending on the application that is made for. Since the focus of this work is on PA applications, drones must meet the following basic specifications: they should be equipped with a high definition (HD) camera for precise monitoring; be capable of lifting a weight of at least 10−12 kg for aerial spraying and be equipped with additional sensors such as compass and GPS module, for efficient and precise navigation and localization.

Unmanned aerial vehicle sensors

In PA applications, the drones employed are semiautonomous as explained in the following. More specifically, a drone maintains its altitude based on a sensor (e.g., an ultrasonic sensor or a barometer or a laser altimeter) and flies at a constant altitude regardless of weather conditions and according to manufacturer specifications. A drone is typically equipped with a global navigation satellite system (GNSS) for precise localization and navigation in global coordinates. The target/destination is given to a drone by either an operator or a mission planner software. Light detection and ranging (LiDAR) is another sensor mounted on UAVs toward detailed 3D laser scanning of the terrain.

Two main routines are running in a drone, as shown in Fig. 6.3 (Daponte et al., 2019). More specifically, drones have a reference generator and an on-board control routine. The waypoint generator takes as input the values x_r, y_r, z_r, ψ_r, x_d, y_d, z_d, ψ_k and u_d, v_d in order to generate the command signal φ_c, Ω_c, ψ_c, θ_c. Then, the on-board control unit takes as input the aforementioned command signal as well as the values P_k, q_k, r_k in order to produce the output signal ω_1, ω_2, ω_3, ω_4 for the motors.

In PA applications, drones are equipped with cameras. A variety of different cameras is available for different types of applications. In particular, red green blue (RGB) cameras are the most popular in PA applications, mainly due to their low cost. Three main technologies of cameras include: thermal, multispectral, and hyperspectral cameras. Table 6.1 summarizes drone cameras for specific PA applications.

Chemical, biological, and meteorological sensors can also be mounted on UAVs to identify chemical and/or organic substances, microorganisms and to monitor weather-related parameters such as temperature and humidity, respectively. Specific types of sensors, in size and weight, are

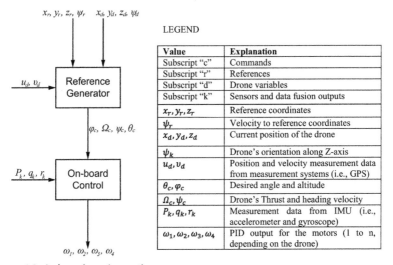

Figure 6.3 A drone's main routines.

Table 6.1 Types of cameras and precision agriculture applications.

Camera	PA application
RGB	Weed mapping
	Terrain mapping
	Orthomosaic
	Digital elevation model
Thermal	Water stress in crops
	Vegetation monitoring
Multispectral and hyperspectral	Leaf water detection
	Ground cover and leaf area index (LAI)
	Normalized difference vegetation index (NDVI)
	Chlorophyll detection
	Leaf water detection
	Vegetation monitoring

selected according to the maximum payload of the UAV of interest. Apart from sensors, a UAV in PA may carry a spraying system, or objects to be transported to specified destinations (Radoglou-Grammatikis et al., 2020). The restriction of payload capacity as well as the extensive use of costly and small platforms may introduce limitations to sensor selection. Note that apart from low weight, sensors must comply with low energy consumption, small size, and high precision (Tsouros et al., 2019b).

Mission planning

For proper use of an airborne imaging system, a variety of different parameters needs to be considered, as explained next. First, the operator must be qualified and well trained, able to recognize all potential threats that might put in danger either the flight or the surroundings. A ground inspection must be done before the flight in the area of interest (AOI) by the operator. Possible threats include electrical cables, high buildings, communication antennas, uneven ground, tall trees, and flock of birds. Considering possible threats, the operator defines safely the drone's operational altitude, the takeoff and landing points, the home point, and an emergency landing point (Sankaran et al., 2015).

A second parameter that needs to be considered is the weather conditions. Shiny weather is considered optimal for UAV applications. Several weather forecast webpages and applications for smart phones and/or tablets exist to provide useful information to the operator; besides wind direction and wind speed. The K-index, which may determine and characterize the magnitude of geomagnetic storms, is also essential. The lower the K-index value (between 0 and 2), the smaller is the interference between ground station and drone. K-index values above 4 are prohibitive for a flight.

Subsequently, the operator must draft a flight plan. This is pursued either by a drone manufacturer application, which is recommended, or by a third-party application. In some cases, a third-party application is the only option, since most of commercial amateur drones have not been developed for mosaicing, therefore they do not support mission planning. For example, the official application for the Mavic Mini from DJI, which is a small budget drone built for entertainment, does not support the development of a flight plan. Well-known mission planning applications include the Foreflight, Drone desk, Litchi, Pix4Dcapture, Drone Harmony, and Drone Deploy.

After all safety measures have been taken and the necessary equipment for the application is determined and mounted on the UAV, the operator has to draft the flight mission. First, if the AOI is excessively large then it needs to be partitioned in smaller, overlapping AOIs followed by computation of partial mission plans, which, properly arranged by the mosaicing program, result in an overall mission plan (Fig. 6.4). In conclusion, for a mission plan, the operator defines the takeoff/landing location, flight altitude, x-overlap, y-overlap, and camera angle.

Mosaicing methods

UAVs acquire data of overlapping images, including front and side overlap, to build a 3D reconstruction model, orthomosaic maps or digital elevation models (DEM) of the field, that include information regarding crop growth. Image mosaicing refers to the combination of images of different angles of overlapping areas, resulting in a high resolution image with georeferenced information (Tahtirvanci and Durdu, 2018). Image mosaicing is necessary when the optical device has a finite field of view, but the desired view is large, as in PA applications that require mapping of several acres of crops. Note that flights from a higher altitude can capture a large-scale image, at the expense of image sharpness. On one hand, UAVs can operate at a higher altitude than traditional land-based surveying methods. On the other hand, UAVs operate on a much lower altitude than satellites. Maps taken from either satellites or from other air vehicles,

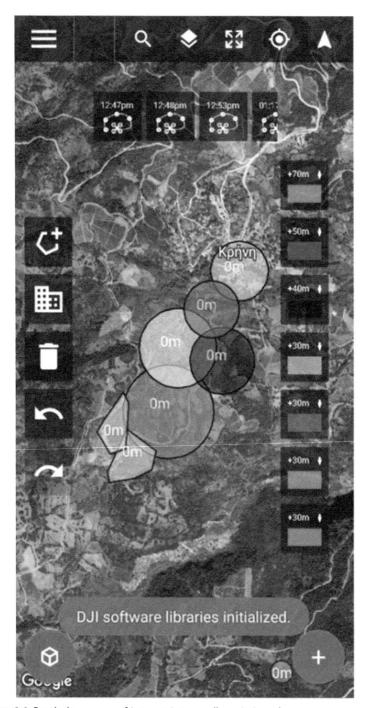

Figure 6.4 Break down area of interest into smaller mission plans.

that is, airplanes, are costly. Moreover, satellite images are not always available for all regions. Since UAVs can fly over a field of interest, they are closer to the ground than either satellites or air vehicles, and therefore, they can supply with higher resolution images.

Photometric techniques refer to the construction of orthomosaic maps and DEMs. Commercial software for mosaicing include the 3DF Zephyr, IMAGINE Photogrammetry, OpenDroneMap, PhotoModeler, Pix4Dmapper, Agisoft PhotoScan, Adobe Photoshop CS6, PixMaker. The evaluation of three different software packages for mosaicing was presented in Song et al. (2016), where Photoshop CC, Autostitch, and Pix4Dmapper were selected for mosaicing airborne images obtained from two different sites. Results showed that Pix4Dmapper is the optimal software when georeferenced images with high accuracy are required. More specifically, Pix4Dmapper resulted in the highest accuracy, 3.31 and 2.61 m, the smallest standard derivation, 2.96 and 1.88 m, and the highest mosaicing speed, 12 and 83 seconds, for sites 1 and 2, respectively.

Image mosaicing consists of five (fine) basic steps, as detailed in the following.

The first step is extraction of features on the images to be mosaicked. Features may be edges, corners, and blobs. In some cases, images are processed by frequency analysis; however, frequency analysis uses all pixel values resulting in long processing times. Real-time applications require reduced computational costs, therefore feature-based methods are then preferable (Ghosh and Kaabouch, 2016). The most well-known algorithms in image mosaicing applications are the scale-invariant feature transform (SIFT) (Lowe, 2004), speeded up robust features (SURF) (Bay et al., 2008), features from accelerated segment test (FAST) (Rosten et al., 2010), and Harris corner detector (Sánchez et al., 2018). However, more algorithms exist such as the binary robust independent elementary features (BRIEF), oriented FAST and rotated (ORB), and other. Feature extraction on UAV images has been carried out in a number of applications, as explained next. In Bi et al. (2014), the scale invariant feature points of remote sensing images from an earthquake region were extracted from UAV images using SIFT. In Liu et al. (2017), SIFT feature extraction was proposed toward image mosaicing of UAV in a road traffic accident scene. SIFT features were extracted in Chen and Huang (2012) for stitching indoor images from a UAV. In Yang et al. (2013), an approach was presented for geometric correction of UAV images without ground control points (GCPs); the SIFT algorithm was used for feature extraction and feature matching. A novel approach was introduced in Yu et al. (2016) that uses UAVs' position information to facilitate a full view of multiple images' stitching. Harris corner detector algorithm was modified to fulfill the real-time requirement of the application. Image features in real-time were extracted by simple descriptors based on the FAST corner detector in Botterill et al. (2010). The method enabled real-time stitching of images captured by an aerial platform. The SURF detector was used due to its robustness in Amiri and Moradi (2016) for feature detection in real-time video mosaicing. A modification of the SURF was introduced in Buyukyazi et al. (2013) for real-time image mosaicing during UAVs surveillance. Extracted feature points are matched with one another, and outliers are removed. Then, homography estimation allows for the images to be aligned. Finally, the images are wrapped and seams are removed by image blending.

The second step is feature matching. Feature matching refers to establishment of correspondence between two images of the same scene. A set of points of interest is extracted and is associated to image descriptors. Then preliminary feature matches between the images are determined. Appropriate detectors and descriptors need to be chosen depending on the application. In aerial images, corner detectors are considered most appropriate for detecting human-made structures from above. Matching algorithms first find a set of distinctive key-points and define regions around them. Then the content of each region is extracted and normalized. Finally, a local descriptor is computed from each normalized region and matching of the local descriptors is carried out. Two well-known feature matching algorithms are the Brute-Force matcher and the fast library for approximate nearest neighbors matcher (FLANN). Feature distance, such as Euclidean distance with Lowe's ratio test, is also used for feature matching (Tahtirvanci and Durdu, 2018).

The third step is the outlier detection and the homography estimation. Acquired images may be subjected to sensory distortions and measurement errors. Therefore outlier features need to be identified and removed. Classic homography algorithms compute the correspondence of image features frame-by-frame using algebraic constraints (Chen et al., 2021a). The quality of homography depends on the number and quality of the extracted features as well as on the employed algorithm. Popular algorithms for homography calculation are the direct linear transform (DLT), different cost functions (algebraic, geometric, reprojection error, Sampson error, etc.), RANdom SAmple Consensus (RANSAC), and least median of squares regression (LMS) (Le et al., 2020).

The fourth step is the image warping. During image warping, the matched key-points of two images are aligned and both images are brought to the same coordinate plane. The process includes the multiplication of the input image with a homography matrix and its warping based on the reference image (Del Gallego and Ilao, 2017). Image warping can be addressed with either parametric (i.e., translation, dilation, rotation, etc.) or nonparametric transformations (i.e., elastic deformations, thin-plate splines, Bayesian approach, etc.).

The fifth step is image blending. Image blending removes illumination differences, seams, movements in the background, sharp intensity changes, etc. by blending overlapped regions of the images. This step is a merging step in which all artifacts in the images are minimized. Image blending methods are partitioned as follows: (1) transition smoothing methods, which are used to provide a nonperceptible transition between images, for example, seam-eliminating function (SEF), Poisson, and gradient-domain image STitching (GIST); (2) optimal seam finding methods, which are used to find the optimal placement for a seam line through an overlapping area, for example, feathering, Dijkstra's algorithm; and (3) hybrid methods, that combine the previous two methods (Prados et al., 2014). Table 6.2 summarizes mosaicing steps and the corresponding algorithms.

Table 6.2 Mosaicing steps and corresponding algorithms.

Mosaicing				
Feature extraction	Feature matching	Homography	Image warping	Image blending
Harris corner	Brute-Force matcher	DLT	Parametric transformations	SEF
SIFT	FLANN	Different cost functions	Nonparametric transformations	GIST
SURF	Feature distance	RANSAC		Feathering
FAST		LMS		Dijkstra's algorithm
BRIEF				Laplacian pyramid
ORB				Poisson

Data processing

Data processing methods analyze the images supplied by UAVs in order to extract crop features, need to process the data in real-time. The latter is necessary for real-time visualization of spatial information acquired from the images in order to support decision-making toward real-time crop management. Several data processing techniques, as well as commercial software, have been developed for fast data processing. Some of the afore-mentioned software packages also offer crop data extraction tools such as Agisoft Photoscan and Pix4Dmapper that can be also used to calculate vegetation indices. Crop features that can be extracted from UAV images are shown in Table 6.3. Information can be extracted for both the crops and the soil by appropriate sensors.

In order to extract information from UAV crop images, the most common crop features calculated are the vegetation indices, which can be calculated separately on each UAV image or to the orthomosaic map depicting the entire crop area. Their values can be used to decide crop characteristics related to biological and/or physical status of crops. Table 6.3 summarizes crop features extracted from UAV images.

Crop information can also be extracted with either data mining techniques or machine learning techniques. Big data collected from the fields related to spectral characteristics of crops can be processed with deep learning to identify crop (Bah et al., 2018; Ferreira et al., 2020; Kitano et al., 2019), to calculate crop growth or crop health (Rasti et al., 2021;

Table 6.3 Crop features extracted from unmanned aerial vehicle images.

Area	Crop feature
Crop	Vegetation indices
	Biomass
	Height
	Size and shape
	Temperature
	Humidity
	Nitrogen
	Color
	Chlorophyll
Soil	Humidity
	Temperature
	Electrical conductivity
	Biomass

Xu et al., 2021b). Crop rows identification with UAVs is mainly used for navigation purposes, that is, to calculate paths for autonomous navigation of agricultural vehicles (Badeka et al., 2020; Bah et al., 2020; Basso and Pignaton de Freitas, 2020).

Unmanned aerial vehicles in precision agriculture

In what follows, the most common applications of drones in PA from 2018 to date are further investigated. References regarding (1) monitoring applications related to crop features are included in Table 6.3 and (2) spraying applications regarding both single UAV and multi-UAV systems are included in Tables 6.4—6.11. All tables display main characteristics as well as applications via which we discuss advantages/disadvantages and outline future research directions.

Single unmanned aerial vehicle systems

In this subsection, we investigate PA applications involving a single UAV. Note that the majority of PA applications involve singe UAVs. However, the power autonomy of a single UAV is usually not enough to cover large agricultural expanses; therefore either multiple successive flights of a single UAV or multi-UAV systems are required, as explained below.

Monitoring

This subsection investigates UAV applications that carry out monitoring flights in PA. Monitoring involves the collection of crop images followed by information extraction toward useful decision-making. Spatial, spectral, and temporal information can be extracted from images, depending on the mounted optical sensor.

Vegetation indices can be calculated from aerial images. Researchers take advantage of the fact that vegetation reflectance changes significantly in different passbands, in order to obtain useful information. More specifically, vegetation indices are computed in different wavelength reflectance by mathematical equations involving the light reflectance of crops in bandwidths, especially in green, red, and near-infrared (NIR). Based on vegetation indices, information regarding leaf-crop health can be induced since the absorption of light wavelengths varies for different leaf compositions. Vegetation indices can be calculated in the visible as well as in the NIR band. In Raeva et al. (2019), multispectral and thermal images were used to derive three vegetation indices that specify a relation between vegetation and soil temperature. Vegetation indices from RGB and multi-spectral images were employed in Du et al. (2018) to estimate the leaf area index (LAI) of rice during its growing period. Four machine learning algorithms were used to estimate LAI: partial least squares regression (PLSR), support vector machine (SVM), random forest (RF), and extreme learning machine (ELM). In Mazzia et al. (2020), multispectral images were used to extract normalized difference vegetation index (NDVI) maps in four different stages of vine growth. A K-means classifier was able to derive vigor maps from the NDVI measurements. Biomass monitoring gave insights regarding pest and weed status, water levels, yield prediction, soil quality, etc. In Devia et al. (2019), multispectral NIR images were used for biomass estimation by calculating seven vegetation indices associated with different stages of rice growth. Table 6.4 summarizes single UAV vegetation indices-based monitoring applications in PA.

Crop growth monitoring applications based on UAV images are also reported. Changes in crops' height provides information regarding the crop health, growth, and reaction to environmental changes. In Ballesteros et al. (2018), onion biomass, in terms of green canopy cover (GCC), canopy height and volume, was estimated from high-resolution RGB images. More specifically, to estimate height and volume of the canopy, a modified version of volume and leaf area index calculation (VOL-LAIC) software was implemented. A real-time phenology detection approach was proposed by Yang et al. (2020) using multispectral and RGB images. A classic phenotypic characteristic is crop height. In Xie et al. (2021), crop height was estimated from UAV images. Crop height monitoring was described in Belton et al. (2019). A photometric crop surface model (CSM) was extracted and compared to the referenced heights of real-time GNSS for validation. Color, texture, and temperature features from both visible and thermal infrared images were used in Liu et al. (2018) toward rice lodging estimation; lodging limits the yield as well as the quality of rice, therefore it needs to be monitored. Structure from motion (SfM) was used in Han et al. (2018) to estimate the height of sorghum plants. The early stage of rice growth was monitored in Rosle et al. (2019); the normalized vegetation index (NDVI) was extracted from UAV images, enabling crop growth observation and damage location in a paddy plot. Monitoring of beet growth was investigated in by Cao et al. (2020); a wide-dynamic-range vegetation index (WDRVI) was calculated, strongly correlated with growing indices such as the fresh weight of leaves (FWL), the fresh weight of roots (FWR),

Table 6.4 Single unmanned aerial vehicle vegetation indices-based monitoring applications in precision agriculture.

References	Scope	Features	Algorithms	Sensors	UAV type	Crop
Raeva et al. (2019)	Soil temperature mapping	Three vegetation indices (NDVI, GNDVI, NDRE) and temperature	Calculation of vegetation indices	Four-band multispectral camera multiSPEC 4C with red, green, red-edge, and NIR bands and a thermal camera thermoMAP	SenseFly fixed-wing drone eBee	Corn and barley
Du et al. (2018)	LAI prediction during growth	Four vegetation indices from multispectral images (NDVI, RVI, NDRE, CI) and three from RGB images (EXG, VDVI, VARI)	Four machine learning methods (PLSR, SVM, RF, ELM)	RGB Sony NEX-7 micro single camera and multispectral camera Ms MQ022MG-CM	Octorotor	Rice
Mazzia et al. (2020)	Vigor maps extraction	Vegetation index (NDVI)	Convolutional neural network (CNN) and a K-means–based classifier	Red and NIR bands of high-resolution multispectral images of UAV camera	UAV airborne Parrot Sequoia	Grapes
Devia et al. (2019)	Biomass estimation	Seven vegetation indices (RVI, NDVI, GNDVI, CTVI, SAVI, DVI, MSAVI)	Multivariable regression model	Tetracam ADC-lite multispectral camera	Quadrotor	Rice

and the LAI. Vegetation monitoring and plant phenotyping were described in Sagan et al. (2019), where three different commercial thermal cameras were tested, mounted on three different UAVs, toward their comparative evaluation in terms of efficacy for plant phenotyping. Table 6.5 summarizes single UAV crop-growth monitoring in PA.

Temperature and humidity are crucial to phenology. UAV remote sensing applications in PA involve extensive thermal as well as humidity information retrieval to describe the timing of certain biological processes.

In Zhang et al. (2018), a method was presented to derive plant temperature from both a UAV thermal infrared camera (TIR) and a charge-couple device (CCD) camera. Temperature data were also derived in Kelly et al. (2019) using an uncalibrated TIR camera. The temperature of a maize canopy was estimated in Zhang et al. (2019b) from high-resolution RGB as well as thermal images toward crop water stress monitoring; four temperature-related indices were used as water stress indicators strongly correlated to the temperature obtained from the leaf stomatal conductance values. Research presented in Ge et al. (2019) investigated soil moisture content estimation using four optimal spectral indices, six methods of pretreatment, and two machine learning algorithms applied on UAV hyperspectral images. The heterogenous areas of irrigated olive groves was identified in Jorge et al. (2019); the proposed photogrammetric method used multispectral images to compute four vegetation indices related to irrigation irregularities. A thermal camera and a multispectral sensor were employed in Chang and Hsu (2018) to acquire data of soil moisture. The soil moisture content was represented by equations to extract relations to indices calculated from the data; temperature vegetation dryness index (TVDI), land surface temperature (LST), and NDVI were calculated. Quantification of water stress was the objective in Das et al. (2021); a thermal remote sensing and machine learning approach was introduced to predict biomass and grain yield of wheat in altering water stress. Five crop water stress-related indices were calculated: standardized canopy temperature index (SCTI), crop water stress index (CWSI), vapor pressure deficit (VPD), crop stress index (CSI), and Jones's thermal index or stomatal conductance index (Ig). In Ihuoma et al. (2021), an approach for irrigation water requirements estimation was presented. Multispectral images acquired from a UAV were used for estimation comparatively to images acquired by two satellite platforms. Evaluation of water status of soybeans was performed in Crusiol et al. (2020). Different water conditions were monitored by TIR UAV imagery

Table 6.5 Single unmanned aerial vehicle crop-growth monitoring applications in precision agriculture.

References	Scope	Features	Algorithms	Sensors	UAV type	Crop
Ballesteros et al. (2018)	Biomass monitoring	Dry leaf biomass (DLB) and dry bulb biomass (DBB), canopy height and volume	VOL–LAIC software in Matlab	RGB PENTAX A40 automatic photography camera	Quadcopter md4–200	Onion
Yang et al. (2020)	Phenology detection	Vegetation index scaled WDRVI, thermal time, leaf development, tillering, flowering, development of fruit, ripening, senescence	CNN, stochastic gradient descent (SGD) optimization algorithm	RedEdge MicaSense multispectral camera and an RGB QX-1 SONY camera	Fixed-wing UAV (MTD-100, CHN)	Rice
Xie et al. (2021)	Crop height estimation	Laser reference height measurements, spectral indices	SfM	RGB camera and a multispectral MicaSense RedEdge 3 camera	DJI Phantom 4 RTK SZ DJI Technology Co.	Inbred rapeseed
Belton et al. (2019)	Crop height	GNSS RTK reference heights	CSM	GoPro 3 + Silver, FinePix X100, Canon SX230HS	In-house build UAV	Wheat
Liu et al. (2018)	Lodging estimation	Color, texture, temperature	Particle swarm optimization and SVM	FLIR thermal camera, a DJI X5R camera, Fluke 941 illuminometer	DJI Inspire 1 drone	Rice

(Continued)

Table 6.5 (Continued)

References	Scope	Features	Algorithms	Sensors	UAV type	Crop
Han et al. (2018)	Crop height estimation	Reference crop height	SfM, multilevel GCPs	Visible-light camera ILCE-6000 Sony	Fixed-wing Tuffwing Mapper	Sorghum
Rosle et al. (2019)	Growth monitoring	Vegetation index (NDVI)	SfM	RGB digital camera	Multirotor UAV	Rice
Cao et al. (2020)	Growth monitoring	WDRVI	Calculation of growing indices (FWR, FWL, LAI, NDVI, WDRVI)	Passive light MicaSence Red Edge-M multispectral optical sensor	DJI PHANTOM4A	Sugar beet
Sagan et al. (2019)	Vegetation monitoring	Phenotyping characteristics (crop height, leaf color, chlorophyll, nitrogen balance index), temperature	LAI-2200C Plant Canopy Analyzer, DUALEX 4 Scientific, SVM classifier	ICI 8640 P-series Infrared Camera, FLIR Vue Pro R 640, thermoMap senseFly	DJI M600 Pro hexacopter, DJI S1000 octocopter, eBee UAV	Soybean and sorghum

toward evaluating comparatively canopy temperature and air temperature. Table 6.6 summarizes single UAV temperature and humidity monitoring applications in PA.

Crop diseases detection is critical in agriculture. Traditionally, disease detection is performed by on-site inspection of crop disease symptoms by agronomists or trained workers. The latter is costly, time consuming, and error-prone due to the subjectivity of the surveillance workers. An accurate and automated identification of crop diseases, including pests and weed, could help developing early responses toward both increasing yield quality and quantity, and reducing economic losses. UAV applications for crop disease detection are common in PA. In Zhang et al. (2019a), a method was proposed for automated crop disease detection using hyperspectral images. A deep convolutional neural network (DCNN) was employed for feature extraction, tested at all stages of the disease spread. Leaf diseases were recognized in Tetila et al. (2020a); four deep learning models were trained with different parameters extracted from UAV images. Identification of infected leaf areas was also demonstrated in Kerkech et al. (2018), where UAV-acquired images from which different vegetation indices, including excess green (ExG), excess red (ExR), excess green—red (ExGR), green—red vegetation index (GRVI), normalized difference indices (NDI), and red—green index (RGI), in three color spaces, namely hue, saturation, value (HSV), LAB, and YUV, were investigated to detect disease symptoms in vineyards. The same research group in Kerkech et al. (2020a) proposed a method for Mildew disease detection; the method was based on visible and infrared images supported by a CNN model. In another work (Kerkech et al., 2020b), the authors tested a new deep learning architecture called vine disease detection network (VddNet), fusing visible, infrared, and depth information. Furthermore, note that weed and pests are responsible for yield loss and deterioration. A classification system for weed identification was proposed in Bah et al. (2019); CNNs were trained using high-resolution UAV images. A method based on CNNs for weed detection was developed in Bah et al. (2018). More specifically, crop rows and interrows were identified first; then, interrow weed was used for training a CNN model. Weed detection was also performed in Kawamura et al. (2021). A commercial RGB camera was employed with the simple linear iterative clustering (SLIC) algorithm and RF classifier. Five deep learning architectures were evaluated in Tetila et al. (2020b) for pests detection (*Spodoptera* spp. and *Anticarsia gemmatalis* species) in soybean fields. Results demonstrated a

Table 6.6 Single unmanned aerial vehicle temperature and humidity monitoring applications in precision agriculture.

References	Scope	Features	Algorithms	Sensors	UAV type	Crop
Zhang et al. (2018)	Temperature estimation	Temperature values of vegetation and soil	OTSU algorithm	FLIR VUE PRO R TIR imager and a DJI Zenmuse Z3 camera	DJI Matrice 600 Pro	Blueberries
Kelly et al. (2019)	Temperature estimation	Temperature values	Linear regression	Nonradiometric TIR Vue Pro 640 camera	Explorian 8 quadcopter Pitchup AB	Mire vegetation
Zhang et al. (2019b)	Water stress monitoring	Temperature index, three water stress indices (DANS, CTSD, CTCV)	Red–green ratio index (RGRI)–Otsu method	FLIR Vue Pro R 640 thermal camera	Hexacopter DJI Phantom 4 Pro	Maize
Ge et al. (2019)	Soil moisture content estimation	Four spectral indices (DI, RI, NDI, PI)	RF and ELM	Headwall Nano-Hyperspec hyperspectral sensor	Six-roto DJI Matrice 600 Pro	Winter wheat
Jorge et al. (2019)	Irrigation irregularities identification	Four vegetation indices (NDVI, GNDVI, SAVI, NDRE)	Correlation analysis	Parrot Sequoia 4.0 multispectral camera	DJI quadcopter Phantom 4 Pro	Olives and grapes
Chang and Hsu (2018)	Soil moisture content estimation	Three moisture content-related indices (TVDI, LST, NVDI)	Linear empirical equations	Zenmuse XT FLIR thermal camera, a Parrot sequoia multispectral sensor and a portable HOBO-U30 weather station	M600 Pro hexacopter	Farm fields

Reference	Application	Parameters	Methods	Sensor	UAV platform	Crop
Das et al. (2021)	Water stress quantification, biomass yield and grain yield estimation	Five water stress indices (SCTI, CWSI, VPD, CSI, Ig)	ENDVI-Otsu method, classification and regression tree (CRT)	High-resolution FLIR Tau 2 longwave thermal infrared (LWIR) camera	DJI Matrice 600 Pro hexacopter	Wheat genotypes
Ihuoma et al. (2021)	Irrigation water requirements estimation	LAI, leaf temperature, NDVI, crop consumptive coefficient, crop evapotranspiration (etc.)	AquaCrop simulation model	Irradiance sensor, Parrot Sequoia multispectral imager	Fixed-wing UAV	Tomatoes
Crusiol et al. (2020)	Water status assessment	Canopy temperature, air temperature, normalized relative canopy temperature (NRCT)	Correlation analysis, regression analysis	thermal infrared DIY-Thermocam camera	Tarot Iron Man 1000 octocopter	Soybean

good generalization capacity of the proposed deep learning algorithms. Table 6.7 summarizes single UAV disease monitoring applications in PA.

Crop ripeness estimation is critical for optimizing farm management as well as for harvesting the desired product quality. Conventional ripeness estimation is accomplished by multiple manual sampling until harvest time, accompanied with chemical analyses, which are not always economically feasible (Vrochidou et al., 2021). Yield estimation also presupposes manual sampling and expert knowledge, which is costly and time consuming in large farms. However, both fast and reliable estimation of agricultural yield are important for efficient crop management. Therefore fast and reliable estimation of both harvest time and agricultural yield before harvest are essential. Currently, remote sensing techniques enable cost-effective as well as nondestructive ripeness and yield estimation. In Zhou et al. (2021), an aerial approach is investigated for maturity of strawberries by classification. A deep learning model, namely You Only Look Once (YOLOv3), was trained from UAV images and classified the maturity in three classes. Coffee ripeness monitoring from UAV images was investigated in Nogueira Martins et al. (2021). A coffee ripeness index (CRI) was extracted, characterized by high sensitivity in ripeness discrimination, compared to traditional vegetation indices. In Apolo-Apolo et al. (2020a), a yield estimation method based on UAV images and deep learning was proposed; a region-based convolutional neural network (RCNN) was trained to detect and count apples. The same research team also applied deep learning for yield and size estimation of citrus (Apolo-Apolo et al., 2020b). Yield estimation of a grapevine was investigated in Di Gennaro et al. (2019); an unsupervised pattern recognition algorithm was employed to estimate cluster number and size from UAV images and, thus, to estimate yield per vine. In Yang et al. (2019), a CNN was trained on UAV images from both RGB and multispectral sensors for rice grain yield estimation. The method was compared to traditional vegetation index-based methods, and promising results were presented. Yield prediction and chlorophyll content estimation was performed in Guo et al. (2020). A new vegetation index, modified red blue VI (MRBVI), was introduced; furthermore, machine learning methods including back propagation neural network model (BP), SVM, RF, ELM were employed to predict yield. In Som-ard et al. (2018), UAV RGB images were fused with ground information for yield estimation of sugarcane. An aerial pumpkin detection and yield estimation method was presented in Wittstruck et al. (2020). Pixel-based image classification, fruit-detection postprocessing

Table 6.7 Single unmanned aerial vehicle disease monitoring applications in precision agriculture.

References	Scope	Features	Algorithms	Sensors	UAV type	Crop
Zhang et al. (2019a)	Yellow rust detection	NDVI, deep features extraction with Inception–Resnet	DCNN	UHD 185 Firefly hyperspectral sensor	DJI S1000 UAV	Winter wheat
Tetila et al. (2020a)	Leaf diseases detection	Deep features extraction	Four deep learning models (VGG-19, ResNet-50, Xception)	1/2.3-in Sony EXMOR sensor	DJI Phantom 3 Professional	Soybean
Kerkech et al. (2018)	ESCA disease detection	Vegetation indices (ExG, ExR, ExGR, GRVI, NDI, RGI)	CNN LeNet-5	RGB sensor	<Not defined >	Grapes
Kerkech et al. (2020a)	Mildew detection	Deep features extraction	CNN SegNet	MAPIR Survey2 RGB camera and infrared light sensor	Scanopy quadcopter	Grapes
Kerkech et al. (2020b)	Mildew detection	Multispectral data and depth map information	CNN VddNet	MAPIR Survey2 RGB camera and infrared light sensor	Quadcopter	Grapes
Bah et al. (2019)	Weed detection	Color vegetation indices (ExG, NDI, CINE, VEG, ExGR)	CNN Alexnet	High-resolution RGB camera	Phantom 3 Pro	Spinach, beet, and bean

(Continued)

Table 6.7 (Continued)

References	Scope	Features	Algorithms	Sensors	UAV type	Crop
Bah et al. (2018)	Weed detection	CIELAB color features, geometric features, edge density, histogram of oriented gradients (HOG), Gabor wavelets, deep features extraction	SLIC, ResNet, SVM, RF	36-Megapixel (MP) RGB camera	DJI Phantom 3 Pro	Spinach, bean
Kawamura et al. (2021)	Weed detection	Canopy height model (CHM), spatial texture, four vegetation indices (Vis, ExG, ExR, GRVI), color index of vegetation extraction (CIVE)	SLIC algorithm and RF classifier	Consumer grade RGB camera	DJI Phantom 4	Rice
Tetila et al. (2020b)	Pests detection	Deep features extraction	SLIC, Inception-v3, Resnet-50, VGG-16, VGG-19, and Xception	1-inch Sony CMOS camera	DJI Phantom 4 Advanced	Soybean

classification, and fruit size and weight quantification were performed. Canopy spectral images were used in Fu et al. (2020) for wheat yield estimation. A number of modeling methods including simple linear regression (LR), multiple linear regression (MLR), stepwise multiple linear regression (SMLR), partial least squares regression (PLSR), artificial neural network (ANN), RF were combined to construct an estimation model. Three cameras were used in Feng et al. (2020): an RGB, a multispectral, and a TIR camera. Eight image features were extracted: NDVI, green normalized difference vegetation index (GNDVI), triangular greenness index (TGI), a channel in CIELAB color space (a*), canopy cover, plant height (PH), canopy temperature, and cotton fiber index (CFI). Models were developed for feature evaluation toward yield estimation. Yield prediction was dealt with in Reza et al. (2019) introducing an alternative method that combined K-means clustering with graph-cut (KCG) algorithm. Segmentation between rice and background was performed with KCG based on color features, and the yield was predicted based on rice area information. Table 6.8 summarizes single UAV yield and ripeness estimation monitoring applications in PA.

Leaf nitrogen concentration (LNC) is also very important for precision management of both irrigation and fertilization. LNC estimation is valuable for guiding nitrogen fertilization. Traditionally, the latter would presuppose plant sampling and laboratory analysis. Remote sensing offers alternative nondistractive solutions. Toward this end, UAV applications in agriculture have been reported, toward LNC estimation. In Lu et al. (2021), the feasibility of an LNC estimation model was evaluated. The model used UAV RGB images of maize field. An estimation approach for LNC in corn was described in Xu et al. (2021a). A set of new vegetation indices has been introduced, and partial least squares (PLS) method was employed to identify optimal relations between them and LNC. A method for LNC estimation regarding citrus was presented in Prado Osco et al. (2019). Spectral vegetation indices have been processed with RF algorithm toward LNC estimation. Estimation of the nitrogen nutrition index (NNI) in rice was performed in Qiu et al. (2021). RGB images were captured during growth of rice and six machine learning algorithms, including adaptive boosting (AB), ANN, K-nearest neighbor (KNN), PLSR, RF, and SVM, were employed for NNI estimation. Table 6.9 summarizes single UAV nitrogen status monitoring applications in PA.

Table 6.8 Single unmanned aerial vehicle yield and ripeness estimation monitoring applications in precision agriculture.

References	Scope	Features	Algorithms	Sensors	UAV type	Crop
Zhou et al. (2021)	Ripeness classification	Deep features	YOLOv3	RGB camera	DJI Phantom 4 Pro	Strawberries
Nogueira Martins et al. (2021)	Ripeness monitoring	CRI ripeness index	Correlation analysis	MicaSense RedEdge MX multispectral camera, RGB camera	DJI Matrice 100, DJI Phantom 4 Pro	Coffee
Apolo–Apolo et al. (2020a)	Yield estimation	Deep features	RCNN	1/2.3" CMOS sensor camera	DJI Phantom 4 Pro	Apples
Apolo–Apolo et al. (2020b)	Yield estimation	Deep features	Faster–R–CNN	RGB Sony Exmor 1/ 2.3" CMOS camera	DJI Phantom 3 Professional	Citrus
Di Gennaro et al. (2019)	Yield estimation	NDVI, linear regression parameters	Unsupervised recognition algorithm	ADC-Snap multispectral camera, Sony Cyber-shot DSC-QX100 RGB camera	Open-source platform with modified multirotor Mikrokopter	Grapes
Yang et al. (2019)	Yield estimation	Deep features, NDVI, LAI	AlexNet	RedEdge MicaSense multispectral camera, a6000 and QX-1 SONY RGB cameras	Fixed-wing (MTD-100)	Rice grain

Reference	Application	Features	Methods	Camera	UAV platform	Crop
Guo et al. (2020)	Yield prediction and chlorophyll estimation	MRBVI	BP, SVM, RF, ELM	High-resolution RGB camera	DJI Phantom 4 Pro V2.0	Maize
Som-ard et al. (2018)	Yield estimation	ExG, leaf size, leaf color, scale parameters, shape, compactness	Object-based image analysis and pixel-based image analysis	RGB camera	eBee Sensefly	Sugarcane
Wittstruck et al. (2020)	Yield estimation	Fruit weight and volume	RF, linear regression	RGB camera	DJI Phantom 4	Hokkaido Pumpkin
Fu et al. (2020)	Yield estimation and growth monitoring	Nine vegetation indices for LAI, leaf dry matter, and yield estimation	LR, MLR, SMLR, PLSR, ANN, RF	Multispectral camera	Six-rotor DJI M600 Pro	Wheat
Feng et al. (2020)	Yield estimation	NDVI, GNDVI, TGI, (a*), PH, canopy temperature, CFI	First order linear model and an intrinsically linear model	XNite Canon Elph130 multispectral camera, ICI 8640P thermal camera, GoPro Hero5 RGB camera	DJI Matrice 600 Pro	Cotton
Reza et al. (2019)	Yield estimation	Color features in lab color space	K-means clustering, KCG algorithm	Sony Alpha 5100 (APS-C)	DJI-1000 rotary-wing-type with eight rotors	Rice

Table 6.9 Single unmanned aerial vehicle nitrogen status monitoring applications in precision agriculture.

References	Scope	Features	Algorithms	Sensors	UAV type	Crop
Lu et al. (2021)	LNC estimation	Four vegetation indices in the visible spectrum, four integrative vegetation indices	Linear models	RGB camera	Six-rotor DJI M600	Summer maize
Xu et al. (2021a)	LNC estimation	Coverage adjusted spectral indices (CASIs)	Random frog algorithm (RFA), PLS	High-definition Cybershot Sony DSC–QX100 camera, Parrot Sequoia multispectral camera	DJI-S1000 +	Corn
Prado Osco et al. (2019)	LNC estimation	Spectral vegetation indices	RF	Parrot Sequoia camera	eBee SenseFly	Citrus
Qiu et al. (2021)	NNI estimation	11 Vegetation indices	AB, ANN, KNN, PLSR, RF, SVM	RGB camera	DJI Phantom 4 Professional	Rice

Spraying

Change of climate within the last decades resulted in outbreaks of pests in farm fields. UAV spraying applications of fertilizers or pesticides offer a sustainable solution for crop health and protection (Chen et al., 2021b). The design of a pesticide sprayer UAV was proposed in Rao and Rao (2019). The system was tested for spraying in a semiautonomous mode. In Meng et al. (2020), a spraying system was investigated regarding droplet distribution on peach trees of various shapes. UAVs for spraying pesticides were designed in Desale et al. (2019) as well as in Garre and Harish (2018); nevertheless, they were not tested in real crop spraying applications. In Abd. Kharim et al. (2019), a standard UAV spraying system available in the market was evaluated on droplet deposition density in a selected area. A UAV system for weeds mapping and site-specific spraying was presented in Hunter et al. (2020). The system was tested in two sod fields reporting promising results. Table 6.10 summarizes single UAV spraying applications in PA.

Multiunmanned aerial vehicle systems

Single UAV systems are characterized by their limited flight duration; therefore, they cannot complete tasks in large fields in a single flight. Furthermore, note that many successive battery charges make various tasks prohibitively long in time. Recently, cooperative robots have been proposed in PA applications (Lytridis et al., 2021). UAV swarms have the potential to increase efficiency in PA. However, multiple UAVs flying at the same time are banned in many countries since local regulations demand a human operator to be fully in charge of a flying UAV all the time. The latter is not possible when many UAVs are flying simultaneously. Despite difficulties, swarm control algorithms for multiple UAVs in agricultural applications have been developed for monitoring and spraying, as described next.

In Ju and Son (2019), multiple UAVs were assumed for agricultural tasks such as monitoring, spraying, or sowing. The authors proposed a distributed swarm control algorithm for multiple UAVs that can be remotely controlled by a single operator. The system was implemented in a simulator to verify the proposed algorithm. A multiagent technology and a prototype multi-UAV system was presented in Skobelev et al. (2018). The prototype was able to connect multiple UAVs in joint survey missions adaptively. Simulation experiments and test flights were conducted to

Table 6.10 Single unmanned aerial vehicle spraying applications in precision agriculture.

References	Scope	Tank capacity	Controller board	UAV type	Crop
Rao and Rao (2019)	Spraying pesticide	4.8 kg	Pixhawk	Four motor in–house UAV	<Not applied >
Meng et al. (2020)	Spraying pesticide	12 L	<Not defined >	Gas-powered helicopter 3WQF120–12	Peaches
Desale et al. (2019)	Spraying pesticide	<Not defined >	KK 2.1.5 controller	Four motor in–house UAV	<Not applied >
Garre and Harish (2018)	Spraying pesticide	5 L	Pixhawk STM32F427	Four motor in–house UAV	<Not applied >
Abd. Kharim et al. (2019)	Spraying organic liquid fertilizer	10 L	<Not defined >	Eight multirotor propeller in–house UAV	Rice
Hunter et al. (2020)	Spraying for weed treatment	<Not defined >	<Not defined >	DJI AGRAS MG-1 octocopter	Sod fields

evaluate the system, for different numbers of agents. In Ju and Son (2018), a multi-UAV system based on the distributed swarm control algorithm was proposed. The system was tested under both autonomous and remote control; furthermore, a comparison to a single UAV system has demonstrated a significantly superior performance. A multi-UAV spraying system was developed in Li et al. (2022). The system was simulated with six UAVs and tested in an actual spraying operation with five UAVs. A grid-based method based on the simulated annealing algorithm was proposed in Apostolidis et al. (2022). Extensive simulated evaluation has been conducted, involving up to 15 UAVs. Moreover, two real-life applications, one in agriculture and one in search and rescue, were demonstrated. Table 6.11 summarizes multi-UAV applications in PA.

Discussion

Agricultural UAVs display a promising potential for eco-friendly and high-efficiency agriculture, in addition to their low maintenance costs. They can adapt to a variety of PA applications; therefore, their use is growing rapidly as an innovative technology for smart farming.

Tables 6.4–6.9 demonstrate that different algorithms are employed, not only in different UAV applications, but also in the same one. In cases like disease monitoring, where the problem is treated by image segmentation, machine learning techniques are dominant due to their effectiveness. However, machine learning algorithms involving many parameters require long training time. Preferably, machine learning algorithms should be endowed with high accuracy with few training parameters and short training time.

In most applications, calculation of vegetation indices is necessary in order to extract characteristics of the crops. UAVs demonstrate great potential in applications related to vegetation indices since UAVs can adopt hyperspectral sensing technology to calculate vegetation indices with accuracy. Temperature and humidity estimation involve thermal sensors for crop or soil monitoring. However, thermal imaging may be prone to errors due to heat emitted from objects, longwave radiation, and environmental conditions. In general, weather conditions play a critical role in the accuracy of sensory measurements by UAVs.

Yield prediction by UAVs is also promising. The latter is attributed to the relation of biomass with the density of production. Biomass can be calculated from aerial images by vegetation indices; therefore, accurate

Table 6.11 Multiunmanned aerial vehicle applications in precision agriculture.

References	Scope	Control inputs	Control	Number of UAVs	UAV type	Crop
Ju and Son (2019)	Swam control algorithm design	Velocity control, formation control, collision avoidance control	Teleoperation with haptic feedback	4	Simulation of quadrotor-type UAVs	<Not applied >
Skobelev et al. (2018)	Prototype multiagent system design	Coordinate flight plan, adaptive reconfigure plan due to disruptions in observation squares	Raspberry Pi 2 and integrated with Pixhawk PX4 flight controller	1, 3, 4, 10	xBee	<Not applied >
Ju and Son (2018)	Multi–UAV system design	Velocity control, formation control, obstacle avoidance control	Autonomous control and teleoperation	3	Quadrotor-type	<Not applied >
Li et al. (2022)	Multi–UAV spraying system	Order irrelevant enumeration solution (OIES) algorithm, equal task assignment (ETA), sequential task assignment (STA), discrete particle swarm optimization (DPSO), and a full permutation algorithm	Equal number of operators through a common application	5, 6	XAircraft XAG P20s and XAG P30	Undefined field
Apostolidis et al. (2022)	Photometry multi–UAV system	Simulated annealing algorithm, DARP algorithm, node placement optimization	Autonomous control	2	DJI Phantom 4 Pro	Undefined field

estimations are computed in most cases. When yield estimation is exclusively based on fruit detection then UAVs often fail because they cannot detect all the fruits on the trees. Note that fruits are visible mainly from underneath a tree rather than from above a tree. This is the main reason why UAV ripeness estimation applications are rather scarce in the literature; remote sensing cannot always detect crop production due to leaf coverage. A solution to the latter limitation could be the adoption of a different crop structure; for instance, different geometries of the foliage could make fruits visible from above. Alternatively, UAVs can fly lower so as to facilitate object detection. Apart from fruits, other objects of interest may be identified by UAVs such as pests, weed, and diseases. Moreover, note that for UAVs that fly near the ground, the environment noise would be significantly decreased, if not eliminated at all. Spraying from low altitude is also beneficial since it is carried out closer to the plant; hence, there is less pesticide waste in the environment.

It appears that UAV spraying applications are not mature yet. Several research studies suggest spraying systems, but only some of them have been tested on-the-field. Arial spraying cannot always be effective due to the limited loading capacity of UAVs since only a small tank can be mounted on board. More powerful UAVs need to be developed to accommodate large payloads for PA tasks. A preferable, integrated spraying system should detect pests/diseases on the field, then decide and carry out with precision spraying on the infected areas. In the aforementioned manner, monitoring and spraying could jointly be carried out by a single UAV in a single mission. UAV designers for PA applications need to consider UAV-customized electronics such as lighter sensors and powerful batteries.

From the literature, it appears that UAVs are used in PA alternatively to satellites since they can comparatively provide high data quality at low costs, for any selected region. However, integration of UAV systems in PA applications involves many challenges; for instance, fields can be vast, UAV platforms are only capable of limited autonomy, limited data storage capacity for high-resolution data, and limited payload for extra sensors or other equipment, for example, for spraying.

Multi-UAV systems are developed to overcome the aforementioned single-UAV limitations. A multi-UAV system can carry out a PA task faster and cover a large agricultural area in a single mission. Future work is needed on multi-UAV control strategies. Moreover, UAVs should be able for real-time data processing and decision-making. The latter would endow

UAVs with capacities to navigate safely, perform obstacle avoidance, adapt to changes, and communicate collaboratively with one another.

Conclusion

The use of UAVs in PA is relatively new and, currently, rather restricted; however, it is expected that it will expand in the future. This work has reviewed the most recent PA applications of UAVs. Main characteristics and main sensors of UAVs were presented; control strategies and data processing methods were also reviewed. Most popular applications in PA, in the last 5 years, were summarized regarding both single UAV and multi-UAV applications. Limitations and promising trends were identified toward improving design in future UAV-based PA applications.

Acknowledgments

We acknowledge support of this work by the project "Technology for Skillful Viniculture (SVtech)" (MIS 5046047), which is implemented under the Action "Reinforcement of the Research and Innovation Infrastructure" funded by the Operational Program "Competitiveness, Entrepreneurship and Innovation" (NSRF 2014−2020) and cofinanced by Greece and the European Union (European Regional Development Fund).

References

Abd. Kharim, M.N., Wayayok, A., Mohamed Shariff, A.R., Abdullah, A.F., Husin, E.M., 2019. Droplet deposition density of organic liquid fertilizer at low altitude UAV aerial spraying in rice cultivation. Comput. Electron. Agric. 167, 105045. Available from: https://doi.org/10.1016/j.compag.2019.105045.

Airforce Technology, 2022. Predator RQ-1 / MQ-1 / MQ-9 Reaper UAV. https://www.airforce-technology.com/projects/predator-uav/.

Amiri, A.J., Moradi, H., 2016. Real-time video stabilization and mosaicking for monitoring and surveillance. In: 2016 4th International Conference on Robotics and Mechatronics (ICROM), 613−618. Available from: https://doi.org/10.1109/ICRoM.2016.7886813.

Apolo-Apolo, O.E., Martínez-Guanter, J., Egea, G., Raja, P., Pérez-Ruiz, M., 2020a. Deep learning techniques for estimation of the yield and size of citrus fruits using a UAV. Eur. J. Agron. 115, 126030. Available from: https://doi.org/10.1016/j.eja.2020.126030.

Apolo-Apolo, O.E., Pérez-Ruiz, M., Martínez-Guanter, J., Valente, J., 2020b. A cloud-based environment for generating yield estimation maps from apple orchards using UAV imagery and a deep learning technique. Front. Plant Sci. 11, 1086. Available from: https://doi.org/10.3389/fpls.2020.01086.

Apostolidis, S.D., Kapoutsis, P.C., Kapoutsis, A.C., Kosmatopoulos, E.B., 2022. Cooperative multi-UAV coverage mission planning platform for remote sensing applications. Auton. Robots. Available from: https://doi.org/10.1007/s10514-021-10028-3.

Badeka, E., Vrochidou, E., Tziridis, K., Nicolaou, A., Papakostas, G.A., Pachidis, T. et al., 2020. Navigation route mapping for harvesting robots in vineyards using UAV-based remote sensing. In: 2020 IEEE 10th International Conference on Intelligent Systems (IS), 171−177. Available from: https://doi.org/10.1109/IS48319.2020.9199958.

Bah, M., Hafiane, A., Canals, R., 2018. Deep learning with unsupervised data labeling for weed detection in line crops in UAV images. Remote Sens. 10 (11), 1690. Available from: https://doi.org/10.3390/rs10111690.

Bah, M.D., Dericquebourg, E., Hafiane, A., Canals, R., 2019. Deep learning based classification system for identifying weeds using high-resolution UAV imagery. Adv. Intell. Syst. Comput. 176−187. Available from: https://doi.org/10.1007/978-3-030-01177-2_13.

Bah, M.D., Hafiane, A., Canals, R., 2020. CRowNet: deep network for crop row detection in UAV images. IEEE Access 8, 5189−5200. Available from: https://doi.org/10.1109/ACCESS.2019.2960873.

Ballesteros, R., Ortega, J.F., Hernandez, D., Moreno, M.A., 2018. Onion biomass monitoring using UAV-based RGB imaging. Precis. Agric. 19 (5), 840−857. Available from: https://doi.org/10.1007/s11119-018-9560-y.

Bange, J., Reuder, J., Platis, A., 2021. Unmanned aircraft systems. Springer Handbooks, pp. 1347−1364. Available from: https://doi.org/10.1007/978-3-030-52171-4_49.

Basso, M., Pignaton de Freitas, E., 2020. A UAV guidance system using crop row detection and line follower algorithms. J. Intell. Robot. Syst. 97 (3−4), 605−621. Available from: https://doi.org/10.1007/s10846-019-01006-0.

Bay, H., Ess, A., Tuytelaars, T., Van Gool, L., 2008. Speeded-up robust features (SURF). Comput. Vis. Image Underst. 110 (3), 346−359. Available from: https://doi.org/10.1016/j.cviu.2007.09.014.

Belton, D., Helmholz, P., Long, J., Zerihun, A., 2019. Crop height monitoring using a consumer-grade camera and UAV technology. PFG − J. Photogramm. Remote Sens. Geoinf. Sci. 87 (5−6), 249−262. Available from: https://doi.org/10.1007/s41064-019-00087-8.

Bi, J., Mao, W., Gong, Y., 2014. Research on image mosaic method of UAV image of earthquake emergency. In: 2014 The Third International Conference on Agro-Geoinformatics, 1−6. Available from: https://doi.org/10.1109/Agro-Geoinformatics.2014.6910665.

Botterill, T., Mills, S., Green, R., 2010. Real-time aerial image mosaicing. In: 2010 25th International Conference of Image and Vision Computing New Zealand, 1−8. Available from: https://doi.org/10.1109/IVCNZ.2010.6148850.

Buyukyazi, T., Bayraktar, S., Lazoglu, I., 2013. Real-time image stabilization and mosaicking by using ground station CPU in UAV surveillance. In: 2013 6th International Conference on Recent Advances in Space Technologies (RAST), 121−126. Available from: https://doi.org/10.1109/RAST.2013.6581183.

Békési, B., Palik, M., Vas, T., Tóth, A.H., 2016. Aviation Safety Aspects of the Use of Unmanned Aerial Vehicles (UAV) (pp. 113−121). Available from: https://doi.org/10.1007/978-3-319-28091-2_10.

Cai, G., Dias, J., Seneviratne, L., 2014. A survey of small-scale unmanned aerial vehicles: recent advances and future development trends. Unmanned Syst. 02 (02), 175−199. Available from: https://doi.org/10.1142/S2301385014300017.

Cao, Y., Li, G.L., Luo, Y.K., Pan, Q., Zhang, S.Y., 2020. Monitoring of sugar beet growth indicators using wide-dynamic-range vegetation index (WDRVI) derived from UAV multispectral images. Comput. Electron. Agric. 171, 105331. Available from: https://doi.org/10.1016/j.compag.2020.105331.

Chang, K.-T., Hsu, W.-L., 2018. Estimating soil moisture content using unmanned aerial vehicles equipped with thermal infrared sensors. In: 2018 IEEE International

Conference on Applied System Invention (ICASI), 168−171. Available from: https://doi.org/10.1109/ICASI.2018.8394559.

Chen, J.H., Huang, C.M., 2012. Image stitching on the unmanned air vehicle in the indoor environment. In: Proceedings of the SICE Annual Conference, 402−406.

Chen, G., Liang, Q., Zhong, W., Gao, X., Cui, F., 2021a. Homography-based measurement of bridge vibration using UAV and DIC method. Measurement 170, 108683. Available from: https://doi.org/10.1016/j.measurement.2020.108683.

Chen, H., Lan, Y., Fritz, B.K., Clint Hoffmann, W., Liu, S., 2021b. Review of agricultural spraying technologies for plant protection using unmanned aerial vehicle (UAV). Int. J. Agric. Biol. Eng. 14 (1), 38−49. Available from: https://doi.org/10.25165/j.ijabe.20211401.5714.

Cisternas, I., Velásquez, I., Caro, A., Rodríguez, A., 2020. Systematic literature review of implementations of precision agriculture. Comput. Electron. Agric. 176, 105626. Available from: https://doi.org/10.1016/j.compag.2020.105626.

Clark, R.M., 2000. Uninhabited Combat Aerial Vehicles: Airpower by the People, for the People, But Not with the People. Air University Press.

Crusiol, L.G.T., Nanni, M.R., Furlanetto, R.H., Sibaldelli, R.N.R., Cezar, E., Mertz-Henning, L.M., et al., 2020. UAV-based thermal imaging in the assessment of water status of soybean plants. Int. J. Remote Sens. 41 (9), 3243−3265. Available from: https://doi.org/10.1080/01431161.2019.1673914.

Daponte, P., De Vito, L., Glielmo, L., Iannelli, L., Liuzza, D., Picariello, F., et al., 2019. A review on the use of drones for precision agriculture. IOP Conf. Ser. 275 (1), 012022. Available from: https://doi.org/10.1088/1755-1315/275/1/012022.

Das, S., Christopher, J., Apan, A., Choudhury, M.R., Chapman, S., Menzies, N.W., et al., 2021. Evaluation of water status of wheat genotypes to aid prediction of yield on sodic soils using UAV-thermal imaging and machine learning. Agric. For. Meteorol. 307, 108477. Available from: https://doi.org/10.1016/j.agrformet.2021.108477.

Desale, R., Chougule, A., Choudhari, M., Borhade, V., Teli, S.N., 2019. Unmanned aerial vehicle for pesticides spraying. IJSART 5, 79−82.

Devia, C.A., Rojas, J.P., Petro, E., Martinez, C., Mondragon, I.F., Patino, D., et al., 2019. High-throughput biomass estimation in rice crops using UAV multispectral imagery. J. Intell. Robot. Syst. 96 (3−4), 573−589. Available from: https://doi.org/10.1007/s10846-019-01001-5.

Du, X., Wan, L., Cen, H., Chen, S., Zhu, J., Wang, H., et al., 2018. Multi-temporal monitoring of leaf area index of rice under different nitrogen treatments using UAV images. Int. J. Precis. Agric. Aviat. 1 (1), 7−12. Available from: https://doi.org/10.33440/j.ijpaa.20200301.57.

Del Gallego, N.P., Ilao, J., 2017. Multiple-image super-resolution on mobile devices: an image warping approach. EURASIP J. Image Video Process. 2017 (1), 8. Available from: https://doi.org/10.1186/s13640-016-0156-z.

Di Gennaro, S.F., Toscano, P., Cinat, P., Berton, A., Matese, A., 2019. A low-cost and unsupervised image recognition methodology for yield estimation in a vineyard. Front. Plant Sci. 10, 559. Available from: https://doi.org/10.3389/fpls.2019.00559.

Feng, A., Zhou, J., Vories, E.D., Sudduth, K.A., Zhang, M., 2020. Yield estimation in cotton using UAV-based multi-sensor imagery. Biosyst. Eng. 193, 101−114. Available from: https://doi.org/10.1016/j.biosystemseng.2020.02.014.

Ferreira, M.P., Almeida, D.R.A.de, Papa, D.de A., Minervino, J.B.S., Veras, H.F.P., et al., 2020. Individual tree detection and species classification of Amazonian palms using UAV images and deep learning. For. Ecol. Manag. 475, 118397. Available from: https://doi.org/10.1016/j.foreco.2020.118397.

Fu, Z., Jiang, J., Gao, Y., Krienke, B., Wang, M., Zhong, K., et al., 2020. Wheat growth monitoring and yield estimation based on multi-rotor unmanned aerial vehicle. Remote Sens. 12 (3), 508. Available from: https://doi.org/10.3390/rs12030508.

Garre, P., Harish, A., 2018. Autonomous agricultural pesticide spraying UAV. In: IOP Conference Series: Materials Science and Engineering, 455, 012030. Available from: https://doi.org/10.1088/1757-899X/455/1/012030.

Ge, X., Wang, J., Ding, J., Cao, X., Zhang, Z., Liu, J., et al., 2019. Combining UAV-based hyperspectral imagery and machine learning algorithms for soil moisture content monitoring. PeerJ 7, e6926. Available from: https://doi.org/10.7717/peerj.6926.

Ghosh, D., Kaabouch, N., 2016. A survey on image mosaicing techniques. J. Vis. Commun. Image Represent. 34, 1–11. Available from: https://doi.org/10.1016/j.jvcir.2015.10.014.

Guettier, C., Sechaud, P., Yelloz, J., Allard, G., Lefebvre, I., Peteuil, P., et al., 2009. Improving tactical capabilities with netcentric systems: the Phoenix'08 experimentation. In: MILCOM 2009 - 2009 IEEE Military Communications Conference, 1–7. Available from: https://doi.org/10.1109/MILCOM.2009.5379771.

Guo, Y., Wang, H., Wu, Z., Wang, S., Sun, H., Senthilnath, J., et al., 2020. Modified red blue vegetation index for chlorophyll estimation and yield prediction of maize from visible images captured by UAV. Sensors 20 (18), 5055. Available from: https://doi.org/10.3390/s20185055.

Gupta, L., Jain, R., Vaszkun, G., 2016. Survey of important issues in UAV communication networks. IEEE Commun. Surv. Tutor. 18 (2), 1123–1152. Available from: https://doi.org/10.1109/COMST.2015.2495297.

Han, X., Thomasson, J.A., Bagnall, G.C., Pugh, N.A., Horne, D.W., Rooney, W.L., et al., 2018. Measurement and calibration of plant-height from fixed-wing UAV images. Sensors 18 (12), 4092. Available from: https://doi.org/10.3390/s18124092.

Hunter, J.E., Gannon, T.W., Richardson, R.J., Yelverton, F.H., Leon, R.G., 2020. Integration of remote-weed mapping and an autonomous spraying unmanned aerial vehicle for site-specific weed management. Pest. Manag. Sci. 76 (4), 1386–1392. Available from: https://doi.org/10.1002/ps.5651.

Ihuoma, S.O., Madramootoo, C.A., Kalacska, M., 2021. Integration of satellite imagery and in situ soil moisture data for estimating irrigation water requirements. Int. J. Appl. Earth Observ. Geoinf. 102, 102396. Available from: https://doi.org/10.1016/j.jag.2021.102396.

Jorge, J., Vallbé, M., Soler, J.A., 2019. Detection of irrigation inhomogeneities in an olive grove using the NDRE vegetation index obtained from UAV images. Eur. J. Remote Sens. 52 (1), 169–177. Available from: https://doi.org/10.1080/22797254.2019.1572459.

Ju, C., Son, H., 2018. Multiple UAV systems for agricultural applications: control, implementation, and evaluation. Electronics 7 (9), 162. Available from: https://doi.org/10.3390/electronics7090162.

Ju, C., Son, H.Il, 2019. A distributed swarm control for an agricultural multiple unmanned aerial vehicle system. In: Proceedings of the Institution of Mechanical Engineers, Part I: Journal of Systems and Control Engineering, 233(10), 1298–1308. Available from: https://doi.org/10.1177/0959651819828460.

Kawamura, K., Asai, H., Yasuda, T., Soisouvanh, P., Phongchanmixay, S., 2021. Discriminating crops/weeds in an upland rice field from UAV images with the SLIC-RF algorithm. Plant Prod. Sci. 24 (2), 198–215. Available from: https://doi.org/10.1080/1343943X.2020.1829490.

Kelly, J., Kljun, N., Olsson, P.-O., Mihai, L., Liljeblad, B., Weslien, P., et al., 2019. Challenges and best practices for deriving temperature data from an uncalibrated UAV thermal infrared camera. Remote Sens. 11 (5), 567. Available from: https://doi.org/10.3390/rs11050567.

Kerkech, M., Hafiane, A., Canals, R., 2018. Deep leaning approach with colorimetric spaces and vegetation indices for vine diseases detection in UAV images. Comput. Electron. Agric. 155, 237−243. Available from: https://doi.org/10.1016/j.compag.2018.10.006.

Kerkech, M., Hafiane, A., Canals, R., 2020a. Vine disease detection in UAV multispectral images using optimized image registration and deep learning segmentation approach. Comput. Electron. Agric. 174, 105446. Available from: https://doi.org/10.1016/j.compag.2020.105446.

Kerkech, M., Hafiane, A., Canals, R., 2020b. VddNet: vine disease detection network based on multispectral images and depth map. Remote Sens. 12 (20), 3305. Available from: https://doi.org/10.3390/rs12203305.

Khuwaja, K.S., Chowdhry, B.S., Khuwaja, K.F., Mihalca, V.O., Ţarcǎ, R.C., 2018. Virtual reality based visualization and training of a quadcopter by using RC remote control transmitter. In: IOP Conference Series: Materials Science and Engineering, 444, 052008. Available from: https://doi.org/10.1088/1757-899X/444/5/052008.

Kitano, B.T., Mendes, C.C.T., Geus, A.R., Oliveira, H.C., Souza, J.R., 2019. Corn plant counting using deep learning and UAV images. IEEE Geosci. Remote Sens. Lett. 1−5. Available from: https://doi.org/10.1109/LGRS.2019.2930549.

Kurogbangkaw, H., 2016. Amazon Prime Air: Drone Delivery Commercial Review. Humor and Tech. https://humortechblog.com/2016/12/amazon-prime-air-drone-delivery-commercial-review.html.

Le, H., Liu, F., Zhang, S., Agarwala, A., 2020. Deep homography estimation for dynamic scenes. In: 2020 IEEE/CVF Conference on Computer Vision and Pattern Recognition (CVPR), 7649−7658. Available from: https://doi.org/10.1109/CVPR42600.2020.00767.

Liu, Y., Bai, B., Zhang, C., 2017. UAV image mosaic for road traffic accident scene. In: 2017 32nd Youth Academic Annual Conference of Chinese Association of Automation (YAC), 1048−1052. Available from: https://doi.org/10.1109/YAC.2017.7967565.

Liu, T., Li, R., Zhong, X., Jiang, M., Jin, X., Zhou, P., et al., 2018. Estimates of rice lodging using indices derived from UAV visible and thermal infrared images. Agric. For. Meteorol. 252, 144−154. Available from: https://doi.org/10.1016/j.agrformet.2018.01.021.

Li, Y., Xu, Y., Xue, X., Liu, X., Liu, X., 2022. Optimal spraying task assignment problem in crop protection with multi-UAV systems and its order irrelevant enumeration solution. Biosyst. Eng. 214, 177−192. Available from: https://doi.org/10.1016/j.biosystemseng.2021.12.018.

Lowe, D.G., 2004. Distinctive image features from scale-invariant keypoints. Int. J. Comput. Vis. 60 (2), 91−110. Available from: https://doi.org/10.1023/B:VISI.0000029664.99615.94.

Lu, J., Cheng, D., Geng, C., Zhang, Z., Xiang, Y., Hu, T., 2021. Combining plant height, canopy coverage and vegetation index from UAV-based RGB images to estimate leaf nitrogen concentration of summer maize. Biosyst. Eng. 202, 42−54. Available from: https://doi.org/10.1016/j.biosystemseng.2020.11.010.

Lytridis, C., Kaburlasos, V.G., Pachidis, T., Manios, M., Vrochidou, E., Kalampokas, T., et al., 2021. An overview of cooperative robotics in agriculture. Agronomy 11 (9), 1818. Available from: https://doi.org/10.3390/agronomy11091818.

Maes, W.H., Steppe, K., 2019. Perspectives for remote sensing with unmanned aerial vehicles in precision agriculture. Trends Plant. Sci. 24 (2), 152−164. Available from: https://doi.org/10.1016/j.tplants.2018.11.007.

Magnussen, Ø., Ottestad, M., Hovland, G., 2015. Multicopter design optimization and validation. Model. Identif. Control 36 (2), 67−79. Available from: https://doi.org/10.4173/mic.2015.2.1.

Martinez Leon, A.S., Rukavitsyn, A.N., Jatsun, S.F., 2021. UAV Airframe Topology Optimization. Lect. Notes Mech. Eng. 338−346. Available from: https://doi.org/10.1007/978-3-030-54814-8_41.

Mazzia, V., Comba, L., Khaliq, A., Chiaberge, M., Gay, P., 2020. UAV and machine learning based refinement of a satellite-driven vegetation index for precision agriculture. Sensors 20 (9), 2530. Available from: https://doi.org/10.3390/s20092530.

Meng, Y., Su, J., Song, J., Chen, W.-H., Lan, Y., 2020. Experimental evaluation of UAV spraying for peach trees of different shapes: Effects of operational parameters on droplet distribution. Comput. Electron. Agric. 170, 105282. Available from: https://doi.org/10.1016/j.compag.2020.105282.

Mfiri, J.T., Treurnicht, J., Engelbrecht, J.A.A., 2016. Automated landing of a tethered quadrotor UAV with constant winching force. In: 2016 Pattern Recognition Association of South Africa and Robotics and Mechatronics International Conference (PRASA-RobMech), 1−6. Available from: https://doi.org/10.1109/RoboMech.2016.7813174.

Mohammad Zain, Z., Mohamad, N., Mohamad Ali, Z., 2002. Redesign of de Bothezat Helicopter : The Way Forward for the Rotorcraft Industry. In: 43rd AIAA/ASME/ASCE/AHS/ASC Structures, Structural Dynamics, and Materials Conference. Available from: https://doi.org/10.2514/6.2002-1736.

Morris, L.V., 2015. On or coming to your campus soon: drones. Innov. High. Educ. 40 (3), 187−188. Available from: https://doi.org/10.1007/s10755-015-9323-x.

Nogueira Martins, R., de Carvalho Pinto, F.de A., Marçal de Queiroz, D., Magalhães Valente, D.S., Fim Rosas, J.T., 2021. A novel vegetation index for coffee ripeness monitoring using aerial imagery. Remote Sens. 13 (2), 263. Available from: https://doi.org/10.3390/rs13020263.

Palik, M., Nagy, M., 2019. Brief history of UAV development. Repüléstudományi Közlemények 31 (1), 155−166. Available from: https://doi.org/10.32560/rk.2019.1.13.

Parihar, P., Bhawsar, P., Hargod, P., 2016. Design & development analysis of quadcopter. Int. J. Adv. Comput. Technol.

Prado Osco, L., Marques Ramos, A.P., Roberto Pereira, D., Akemi Saito Moriya, É., Nobuhiro Imai, N., Takashi Matsubara, E., et al., 2019. Predicting canopy nitrogen content in citrus-trees using random forest algorithm associated to spectral vegetation indices from UAV-imagery. Remote Sens. 11 (24), 2925. Available from: https://doi.org/10.3390/rs11242925.

Prados, R., Garcia, R., Neumann, L., 2014. Image blending techniques and their application in underwater mosaicing. Found. Trends Comput. Graph. Vis. 13, 192.

Qiu, Z., Ma, F., Li, Z., Xu, X., Ge, H., Du, C., 2021. Estimation of nitrogen nutrition index in rice from UAV RGB images coupled with machine learning algorithms. Comput. Electron. Agric. 189, 106421. Available from: https://doi.org/10.1016/j.compag.2021.106421.

Radoglou-Grammatikis, P., Sarigiannidis, P., Lagkas, T., Moscholios, I., 2020. A compilation of UAV applications for precision agriculture. Comput. Netw. 172, 107148. Available from: https://doi.org/10.1016/j.comnet.2020.107148.

Raeva, P.L., Šedina, J., Dlesk, J., 2019. Monitoring of crop fields using multispectral and thermal imagery from UAV. Eur. J. Remote Sens. 52 (sup1), 192−201. Available from: https://doi.org/10.1080/22797254.2018.1527661.

Rao, V.P.S., Rao, G.S., 2019. Design and modelling of an affordable UAV based pesticide sprayer in agriculture applications. In: 2019 Fifth International Conference on

Electrical Energy Systems (ICEES), 1−4. Available from: https://doi.org/10.1109/ ICEES.2019.8719237.

Rasti, S., Bleakley, C.J., Silvestre, G.C.M., Holden, N.M., Langton, D., O'Hare, G.M.P., 2021. Crop growth stage estimation prior to canopy closure using deep learning algorithms. Neural Comput. Appl. 33 (5), 1733−1743. Available from: https://doi.org/ 10.1007/s00521-020-05064-6.

Reza, M.N., Na, I.S., Baek, S.W., Lee, K.-H., 2019. Rice yield estimation based on K-means clustering with graph-cut segmentation using low-altitude UAV images. Biosyst. Eng. 177, 109−121. Available from: https://doi.org/10.1016/j.biosystemseng. 2018.09.014.

Rosle, R., Che'Ya, N., Roslin, N., Halip, R., Ismail, M., 2019. Monitoring early stage of rice crops growth using normalized difference vegetation index generated from UAV. IOP Conf. Ser. 355 (1), 012066. Available from: https://doi.org/10.1088/1755-1315/ 355/1/012066.

Rosten, E., Porter, R., Drummond, T., 2010. Faster and better: a machine learning approach to corner detection. IEEE Trans. Pattern Anal. Mach. Intell. 32 (1), 105−119. Available from: https://doi.org/10.1109/TPAMI.2008.275.

Sagan, V., Maimaitijiang, M., Sidike, P., Eblimit, K., Peterson, K., Hartling, S., et al., 2019. UAV-based high resolution thermal imaging for vegetation monitoring, and plant phenotyping using ICI 8640 P, FLIR Vue Pro R 640, and thermoMap cameras. Remote Sens. 11 (3), 330. Available from: https://doi.org/10.3390/rs11030330.

Sankaran, S., Khot, L.R., Espinoza, C.Z., Jarolmasjed, S., Sathuvalli, V.R., Vandemark, G.J., et al., 2015. Low-altitude, high-resolution aerial imaging systems for row and field crop phenotyping: a review. Eur. J. Agron. 70, 112−123. Available from: https://doi.org/10.1016/j.eja.2015.07.004.

Skobelev, P., Budaev, D., Gusev, N., Voschuk, G., 2018. Designing multi-agent swarm of UAV for precise agriculture. Commun. Comput. Inf. Sci. 47−59. Available from: https://doi.org/10.1007/978-3-319-94779-2_5.

Som-ard, J., Hossain, M.D., Ninsawat, S., Veerachitt, V., 2018. Pre-harvest sugarcane yield estimation using UAV-based RGB images and ground observation. Sugar Tech. 20 (6), 645−657. Available from: https://doi.org/10.1007/s12355-018-0601-7.

Song, H., Yang, C., Zhang, J., Hoffmann, W.C., He, D., Thomasson, J.A., 2016. Comparison of mosaicking techniques for airborne images from consumer-grade cameras. J. Appl. Remote Sens. 10 (1), 016030. Available from: https://doi.org/10.1117/ 1.JRS.10.016030.

Sánchez, J., Monzón, N., Salgado, A., 2018. An analysis and implementation of the Harris corner detector. Image Process. Line 8, 305−328. Available from: https://doi.org/ 10.5201/ipol.2018.229.

Tahtirvanci, A., Durdu, A., 2018. Performance analysis of image mosaicing methods for unmanned aerial vehicles. In: 2018 10th International Conference on Electronics, Computers and Artificial Intelligence (ECAI), 1−7. Available from: https://doi.org/ 10.1109/ECAI.2018.8679007.

Tetila, E.C., Machado, B.B., Astolfi, G., Belete, N.A., de, S., Amorim, W.P., et al., 2020a. Detection and classification of soybean pests using deep learning with UAV images. Comput.s Electron. Agric. 179, 105836. Available from: https://doi.org/ 10.1016/j.compag.2020.105836.

Tetila, E.C., Machado, B.B., Menezes, G.K., Da Silva Oliveira, A., Alvarez, M., Amorim, W.P., et al., 2020b. Automatic recognition of soybean leaf diseases using UAV images and deep convolutional neural networks. IEEE Geosci. Remote Sens. Lett. 17 (5), 903−907. Available from: https://doi.org/10.1109/LGRS.2019. 2932385.

Tsouros, D.C., Bibi, S., Sarigiannidis, P.G., 2019a. A review on UAV-based applications for precision agriculture. Information 10 (11), 349. Available from: https://doi.org/10.3390/info10110349.

Tsouros, D.C., Triantafyllou, A., Bibi, S., Sarigannidis, P.G., 2019b. Data acquisition and analysis methods in UAV- based applications for precision agriculture. In: 2019 15th International Conference on Distributed Computing in Sensor Systems (DCOSS), 377−384. Available from: https://doi.org/10.1109/DCOSS.2019.00080.

Vrochidou, E., Bazinas, C., Manios, M., Papakostas, G.A., Pachidis, T.P., Kaburlasos, V.G., 2021. Machine vision for ripeness estimation in viticulture automation. Horticulturae 7 (9), 282. Available from: https://doi.org/10.3390/horticulturae7090282.

Wikipedia, 2023a. Ryan Firebee. https://en.wikipedia.org/wiki/Ryan_Firebee#/media/File:BQM-34F_launch_Tyndall_AFB_1982.JPEG.

Wikipedia, 2023b. Target Drone. https://en.wikipedia.org/wiki/Target_drone#/media/File:Winston_Churchill_and_the_Secretary_of_State_for_War_waiting_to_see_the_launch_of_a_de_Havilland_Queen_Bee_radio-controlled_target_drone,_6_June_1941._H10307.jpg.

Wittstruck, L., Kühling, I., Trautz, D., Kohlbrecher, M., Jarmer, T., 2020. UAV-based RGB imagery for Hokkaido Pumpkin (Cucurbita max.) detection and yield estimation. Sensors 21 (1), 118. Available from: https://doi.org/10.3390/s21010118.

Xie, T., Li, J., Yang, C., Jiang, Z., Chen, Y., Guo, L., et al., 2021. Crop height estimation based on UAV images: methods, errors, and strategies. Comput. Electron. Agric. 185, 106155. Available from: https://doi.org/10.1016/j.compag.2021.106155.

Xu, X., Fan, L., Li, Z., Meng, Y., Feng, H., Yang, H., et al., 2021a. Estimating leaf nitrogen content in corn based on information fusion of multiple-sensor imagery from UAV. Remote Sens. 13 (3), 340. Available from: https://doi.org/10.3390/rs13030340.

Xu, J., Yang, J., Xiong, X., Li, H., Huang, J., Ting, K.C., et al., 2021b. Towards interpreting multi-temporal deep learning models in crop mapping. Remote Sens. Environ. 264, 112599. Available from: https://doi.org/10.1016/j.rse.2021.112599.

Yang, Y., Sun, G., Zhao, D., Peng, B., 2013. A real time mosaic method for remote sensing video images from UAV. J. Signal. Inf. Process. 04 (03), 168−172. Available from: https://doi.org/10.4236/jsip.2013.43B030.

Yang, Q., Shi, L., Han, J., Zha, Y., Zhu, P., 2019. Deep convolutional neural networks for rice grain yield estimation at the ripening stage using UAV-based remotely sensed images. Field Crop. Res. 235, 142−153. Available from: https://doi.org/10.1016/j.fcr.2019.02.022.

Yang, Q., Shi, L., Han, J., Yu, J., Huang, K., 2020. A near real-time deep learning approach for detecting rice phenology based on UAV images. Agric. For. Meteorol. 287, 107938. Available from: https://doi.org/10.1016/j.agrformet.2020.107938.

Yan, C., Fu, L., Zhang, J., Wang, J., 2019. A comprehensive survey on UAV communication channel modeling. IEEE Access 7, 107769−107792. Available from: https://doi.org/10.1109/ACCESS.2019.2933173.

Yu, C., Wang, J., Ding, Y., Shan, J., Xin, M., 2016. Feedback-control-aided image stitching using multi-UAV platform. In: 2016 12th World Congress on Intelligent Control and Automation (WCICA), 2420−2425. Available from: https://doi.org/10.1109/WCICA.2016.7578738.

Zhang, Y., Zhou, J., Meng, L., Li, M., Ding, L. & Ma, J. (2018). A method for deriving plant temperature from UAV TIR image. In: 2018 7th International Conference on Agro-Geoinformatics (Agro-Geoinformatics), 1−5. Available from: https://doi.org/10.1109/Agro-Geoinformatics.2018.8475995.

Zhang, X., Han, L., Dong, Y., Shi, Y., Huang, W., Han, L., et al., 2019a. A deep learning-based approach for automated yellow rust disease detection from high-resolution hyperspectral UAV images. Remote Sens. 11 (13), 1554. Available from: https://doi.org/10.3390/rs11131554.

Zhang, L., Niu, Y., Zhang, H., Han, W., Li, G., Tang, J., et al., 2019b. Maize canopy temperature extracted from UAV thermal and RGB imagery and its application in water stress monitoring. Front. Plant Sci. 10, 1270. Available from: https://doi.org/10.3389/fpls.2019.01270.

Zhou, X., Lee, W.S., Ampatzidis, Y., Chen, Y., Peres, N., Fraisse, C., 2021. Strawberry maturity classification from UAV and near-ground imaging using deep learning. Smart Agric. Technol. 1, 100001. Available from: https://doi.org/10.1016/j.atech.2021.100001.

CHAPTER 7

Drones as functional parts of physical-cyber eco-systems

Aristotelis C. Tagarakis[1], Lefteris Benos[1], Dimitrios Kateris[1], George Kyriakarakos[2] and Dionysis Bochtis[1]
[1]Institute for Bio-Economy and Agri-Technology (IBO), Centre for Research and Technology Hellas (CERTH), Thessaloniki, Greece
[2]farmB Digital Agriculture S.A., Thessaloniki, Greece

Introduction

Without a doubt, agriculture plays a strategic role toward addressing a plethora of challenges, including the need for securing food for an ever-increasing global population (Crist et al., 2022). Food security is a multifaceted goal that seeks to alleviate hunger while guaranteeing a sustainable food supply, based on four intrinsic pillars: (1) food availability; (2) food access; (3) food stability; and (4) food utilization (Abbasi et al., 2022). Additional issues to be addressed stem from climate change (Hasegawa et al., 2021; Agrimonti et al., 2021) affecting several aspects, such as soil degradation and water resources, large-scale changes of weather patterns, including the existence of floods and droughts (Sekhri, 2022; Jhariya et al., 2021; Lampridi et al., 2020; Lampridi et al., 2019a). In a nutshell, as a means of facing the above concerns, there is an urgent necessity to maximize the efficiency of agricultural practices while, simultaneously, limiting the environmental burden.

Following the fourth industrial revolution, which is reshaping every sector, the so-called Agriculture 4.0 has emerged taking advantage of the advancement in information and communication technology (ICT) (Benos et al., 2022a; Liu et al., 2021; Sørensen et al., 2010). Thus, ICT is considered prerequisite of contemporary agriculture. ICT in agriculture indicatively consists of sensors, wireless sensor networks (WSNs), low-cost satellites, agri-robots, online services, cloud analytics, traceability systems, farm management information systems, machine learning (ML), blockchain, as well as unmanned ground and aerial vehicles (Sørensen et al., 2019; Tagarakis et al., 2021; Benos et al., 2022b; Zhai et al., 2020; Rejeb et al., 2022; Benos et al., 2021; Alobid et al., 2022). The large volume of

Unmanned Aerial Systems in Agriculture
DOI: https://doi.org/10.1016/B978-0-323-91940-1.00007-4

data, coming from digital technologies (usually referred to as "big data"), includes, for example, data measured from the field, information on weather, environmental conditions, pH, and soil types. Overall, the challenge with big data concerns the size of the computing and network infrastructure that is required to create a big data facility (Osinga et al., 2022; Kalyani and Collier, 2021; Liakos et al., 2018). Furthermore, proper analyzing, editing, and interpreting, as well as heavy demands on the underlying processing power, are required to turn these data into useful information for site-specific management. To that end, as workloads and analytics technologies move to the "cloud," the ability to perform complex data analysis on very massive datasets has been rapidly increased.

Among the aforementioned cutting-edge technologies supporting modern agriculture, this chapter focuses on unmanned aerial vehicles (UAVs), popularly known as drones. In particular, it elaborates on the usage of UAVs for addressing multiple purposes in the sectors of open-field agriculture, forestry, and livestock farming. A UAV can be simply defined as an aerial robot that can be operated remotely or fly with varying degrees of autonomy based on flight plans and potentially on-board artificial intelligence (AI) systems. UAVs utilize various cameras and sensors, which provide them with notable observation-making competences for a variety of operations. Several classifications have been proposed in the relevant literature (Hassanalian and Abdelkefi, 2017), based on weight and size as well as drone aerodynamic features. In brief, drones vary in size and weight, from huge to insect-like drones, depending on the specific mission requirements. Moreover, they may have fixed or flapping wings and one (monocopters) or several rotors (e.g., quadcopters, hexacopters, octocopters, named based on the number of rotors they use). They can also combine characteristics of these categories (hybrid drones). In general, fixed-wing drones are used for covering large areas at high speed, without, however, being able to hover, a feature making rotary-wing drones excellent aerial vehicles for the inspection of difficult-to-reach areas. Nevertheless, fixed-wing drones are superior in terms of overall flight times and energy consumption. In reference to their application, UAVs can be roughly classified into aerial vehicles aiming to either perceive or act, or carry out both (Maghazei and Netland, 2020; Labib et al., 2021).

This chapter briefly describes the technical aspects of the Horizon Europe research and innovation action project "Multi-purpose physical-cyber agri-forest drones ecosystem for governance and environmental observation—SPADE," which intends to exploit the potential benefits to be reaped from the use of agri-forest UAVs. The ultimate goal is, firstly, to optimize production and, secondly, to facilitate the additional use of the data for applications related to governance and environmental observation.

An overview of unmanned aerial vehicles used in livestock farming, open-field agriculture, and forestry

In general, remote sensing provides a number of techniques for measuring different Earth physical properties by utilizing reflected or emitted energy, at a specific time or period. It has been affected, among others, by the considerable progress in various technologies like global navigation satellite systems and geographical information systems. Remote sensing has long been regarded as a high potential tool and gone hand-in-hand with livestock farming, open-field agriculture, and forestry aiming at optimizing the corresponding management and practices. Remarkably, remote sensing provides an alternative to the limitations of terrestrial Internet of things (IoT) infrastructure and, consequently, new opportunities for society. Human-crewed aircrafts, UAVs, and satellites constitute common platforms facilitating remote sensing. As far as the management of forests and crop and livestock resources is concerned, there exist some advantages rendering UAVs the most effective option to be considered. In particular, satellite imagery with moderate resolution has demonstrated limitations in some applications making them not suitable for on-site agricultural and environmental monitoring purposes (Padró et al., 2019; Khaliq et al., 2019). One of the main advantages of UAVs over satellites is that they can deliver images of high quality and resolution even on cloudy days. Interestingly, fusion of data coming from high-resolution drone datasets and satellite imagery (Zhao et al., 2019; Dash et al., 2018) or ground WSNs and IoT (Popescu et al., 2020) turned out to be a powerful and effective solution for monitoring and decision-making. Moreover, as compared to aircrafts with a human crew, UAVs are more cost-efficient, easier to set up and maintain, and can reduce safety risks (Radoglou-Grammatikis et al., 2020; Tsouros et al., 2019), so that accident-free workspaces are given for farmers (Benos et al., 2020a; Benos and Bochtis, 2021). Key factors for the radical evolution of UAVs include significant advances in sensor technologies as well as improved battery performance and falling prices.

Livestock farming

Livestock refers to animals which are domesticated for production of milk, eggs, meat, and their by-products. The fundamental element of livestock management is to ensure that animals are properly cared for and fed. Additionally, they should be properly managed to yield a profit. Hence, having appropriate strategies in place is as critical as having effective management systems in place, as this will generate a steady stream of income. Broadly speaking, livestock management requires time and high investment. In the direction of decreasing investment capital and increasing profit, a number of technologies are being used in the framework of precision livestock farming (Schillings et al., 2021; Aquilani et al., 2022). Precision livestock farming was built upon providing better information to the livestock farmer on the animals and utilizing the known principles of process engineering to achieve a level of automation.

One such example of time-saving and cost-efficient technology that has attracted the interest in livestock farming is using UAVs (Benalaya et al., 2022; Vrchota et al., 2022). Such technology has the potential to optimize the effort of a farmer by helping him to seamlessly monitor all the ground, without the burden of manual inspection. Indicatively, UAVs can contribute, to a great extent, to detecting, counting, monitoring, and tracking livestock animals, searching pastures, reporting any abnormal condition, and securing the herds against potential threats (Alanezi et al., 2022). Besides, visual inspection of livestock can help detect potentially sick animals showing symptoms of disease and is regularly applied in traditional practice by the livestock farmers. Visual imaging inspection using UAVs can provide a powerful tool to farmers, providing the ability of more regular and less time-consuming and accurate inspections. The ability to do this in an effective and timely manner allows for faster treatment with a higher likelihood of success and is instrumental in avoiding further infection in flocks and herds. UAV-enabled systems can also measure vital information including size, weight, color, and physical activity. Drones can even record or play sounds when needed, such as a dog's bark, for moving stock, sometimes more efficiently than a dog. New Zealand has been one of the first countries to embrace UAVs for livestock management (Agriland Team, 2019).

Open-field agriculture

In traditional agriculture, farmers and agronomists perform manual scouting of the fields by walking and visually inspecting the plants and soil properties (Gemtos et al., 2013; Anagnostis et al., 2021). However, manual visual inspections of crops and fields are extremely time-consuming and rely on the observer's ability to spot the signs of potential problems. Hence, this method is particularly inefficient for large-scale farms, as it requires a relatively large amount of time. Human observation is also unable to provide real-time monitoring of a farm and could only provide monitoring information for part of a farm on any given day.

Numerous sensors, dedicated to monitor field properties, have been developed and are commercially available providing the possibility to monitor and manage field spatial and temporal variability (Gemtos et al., 2013; Anagnostis et al., 2021). These technological developments combined with the rise of AI and digitization of agriculture, the use of UAV-enabled systems is gaining popularity (Tagarakis et al., 2022). UAVs are capable of carrying out a large range of tasks in real time, using a combination of RGB (red, green, blue) cameras, multispectral imaging, and thermal imaging (Singh et al., 2022). Equipped with the right sensors, a drone can fly above a field and provide the farmer with a detailed report covering everything from crop counts to health conditions. Unmanned aerial systems are used in a plethora of applications. First, UAVs are ideal when it comes to monitoring and tracking, since they can quickly cover the field to monitor the health and growth of crops and soil. This is the primary application of drones because their sensors are capable of tracking the absorption of a specific wavelength, producing a color-contrast image that visually reflects the potential problematic regions. Second, UAVs can be equipped with spectroscopy and thermography sensors which can locate sites where the plants are stressed indicating the need for irrigation. Maintaining crop health can also be aided by spreading fertilizers and pesticides from drones. Another application of drones is their use for surveys, as they can cover large regions that have been damaged by natural disasters in an attempt to determine the underlying causes and consequences of the incidents. The quick response coupled with high-resolution imagery enables information to be collected at large scales, something that is very challenging to be achieved on the same timescale, if manual work is involved (Aslan et al., 2022; Benos et al., 2020b).

In summary, the most prevalent applications performed with UAVs regarding open-field agriculture are monitoring and spraying (Kim et al., 2019). In monitoring operations, specific information on crop, vegetation indicators, and soil parameters is acquired by image analysis and remote sensing (Anagnostis et al., 2021; Tagarakis et al., 2022; Angelopoulou et al., 2019). Consequently, several diseases or pests as well as crop health or growth can be monitored. Concerning spraying applications, the required quantities of fertilizers and pesticides are sprayed to enhance crop yield and prevent diseases. In addition, mapping, irrigation, and weed detection have also been implemented with UAVs (Tagarakis et al., 2022; Dian Bah et al., 2018). It is also worth noting that UAVs can play an important role in cooperating operations with unmanned ground vehicles (Katikaridis et al., 2022). The future of UAVs in cropping is also very promising, since unmanned aerial systems are being developed to serve as mechanical pollinators and incorporate smart applications. These applications are anticipated to render UAVs an affordable technology to address the challenge of fighting food insecurity, however, in a cost-effective manner as in all cases of new technology implementation (Mazinani et al., 2023; Wang et al., 2022; Lampridi et al., 2019b). The aspect of efficient mission planning must be also considered for a cost-effective implementation of UAVs in open-field applications under the consideration that UAVs regard autonomous robotic platforms (Dolias et al., 2022; Moysiadis et al., 2020). Certainly, more progress is needed in this area, but researchers are optimistic that UAVs will be able to effectively transport and disperse pollen seeds in fields. Moreover, UAVs are expected to be used for purposes such as planting, transportation, and field surveys in the near future (Ju and Son, 2018; Marinoudi et al., 2021).

Forestry

Forests account for 31% of the world's terrestrial biodiversity (European Commission, 2022). Forestry constitutes a sector that has long acknowledged the importance of aerial photography, topographic monitoring, and aerial surveillance. As a matter of fact, foresters were among the first to deploy manned air vehicles to carry out aerial surveys to effectively map and monitor vast tracts of forest land (Cromwell et al., 2021). As a consequence, foresters have embraced the use of UAVs for forest management, since the latter can increase the logistical efficiency of getting data in due time. Additionally, UAVs can help in biodiversity and wildlife conservation. As compared with the other traditional remote sensing technologies, unmanned aerial systems are capable for real-time applications, since they combine both high spatial resolution and fast turnaround times by means of lower operational costs. These aerial systems are also efficient in collecting data concerning tree characteristics such as age, diameter, height, and location. This information is particularly valuable when it comes to monitoring the forest's health and taking informed decisions about its management. Generally, the most popular UAVs used in forest management are multicopters, even though fixed-wing drones remain the preferred choice for mapping or surveying large forest areas.

UAVs can carry several sensors to be task-oriented. Indicatively, thermal sensors can be adapted to UAVs with the intention of detecting fires in forests, thus, contributing to control the spread of the fire in space and time (Torresan et al., 2017). Identifying the burned surface area is very challenging, as it may look like being unburned in imagery provided by satellites or aircrafts. To that end, Shin et al. (2019) examined the potential of multispectral images taken by UAVs that can be applied for the classification of burn severity. This method, through combination of multispectral images from UAVs and supervised classifiers, proved to be sufficiently accurate. In Fraser and Congalton (2019), the efficacy of UAVs was demonstrated for collecting thematic map accuracy evaluation reference data in the forests of New England, in the United States. Furthermore, in Qiu et al. (2018), a UAV photogrammetric survey system was found to be suitable for recording ancient tree communities. Also, Feduck et al. (2018) examined the potential of utilizing drones to quickly detect conifer seedlings in Canada. To conclude, as stated in Matese (2020), the primary applications of the forestry sector are directed toward fire monitoring, disease mapping, species classification, inventory resource, and spatial gap assessment. Finally, UAVs can be used in surveillance and security operations, monitoring illegal activities, such as unlicensed logging, locating missing persons, and locating invasive species.

The concept of a multipurpose physical-cyber agri-forest drones ecosystem for governance and environmental observation

In this section, the concept for the development of an intelligent ecosystem is presented in order to address the multiple purposes concept in the light of deploying UAVs to promote sustainable digital services for the benefit of a large scope of end-users in the sectors of crop production, forestry, and livestock; however, more use cases may be applicable (Fig. 7.1). The ecosystem includes individual UAV usability, UAV type applicability (e.g., swarm, collaborative, autonomous, tethered), UAV governance models availability, and UAV-generated data trustworthiness. Multipurpose is further determined in the sensing data space reusability of trained AI/ML models. This enables sustainability and resilience of the overall life cycle of

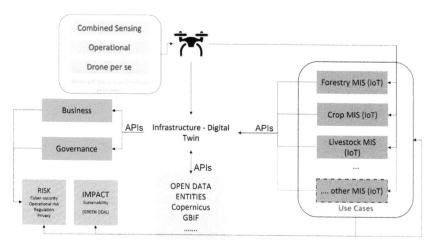

Figure 7.1 The conceptual framework of the proposed ecosystem.

developing, setting up, offering, providing, testing, validating, refining as well as enhancing digital transformations and "innovation building" services in livestock farming, open-field agriculture, and forestry.

In this context, a digital platform is proposed that is able to realize the potential benefits to be reaped from use of drones. This platform is making drone operations better accessible providing a service channel for value-added services (business and governance) enabled by drones. Within this concept, three main domains are addressed as use cases that are benefitted by the drone ecosystem; crop production, livestock, and forestry. While demonstrating the use cases, the benefits coming from the use of drones are analyzed and quantified, on a detailed stakeholder-level basis. The demonstrations serve as an analysis platform to investigate the regulatory framework at international and national level. A thorough stakeholder analysis is needed, detailing the risks, costs, and benefits identified in use cases. Also, a detailed risk analysis should be performed, first as desk research and then observing the experiences of dedicated use cases within pilots. The European Union Aviation Safety Agency (EASA) risk framework should be thoroughly reviewed to identify the development needs are identified and reported.

To accomplish the vision of the proposed concept, the following specific objectives have been set:

• Learning and engaging others in learning about the various impacts and risks of drones as multipurpose vehicles.

• Developing cost-effective drones and sensors for multipurpose applications.

- Developing an open-source, open-access ecosystem which allows the collection, processing, and distribution of data from and to multiple stakeholders.
- Making the innovation known, used, socially robust, and anchored for long term in agriculture and forestry.

The abovementioned concept is being materialized in the framework of a European initiative through an EU-funded project named SPADE. Within this project, three pilots are set up incorporating unmanned aerial system (UAS)-specific use cases, testing the applicability of different types of UASs in the sectors of open-field agriculture, forestry, and livestock. The scope of the pilots is to develop and validate the functionalities of the proposed ecosystem, after which a generalization is made to apply the analysis results on a national and European level.

Ground-breaking nature of concept and approach

This chapter provides the concept and the design of a digital platform for effective, accessible, and secure drone operations and services. The added value of this platform is realized through demonstrated user benefits and reduced risks in drone operations, following EASA guidelines and practices, and open-source principles. Several vendor-specific and context-sensitive ICT platforms exist for drone operations and deployment, making them candidates for international standardization of application interfaces and data harmonization.

AI can support more informed and timely decision-making requiring minimum actions by the users. Stakeholders, such as farmers and foresters, do not have the time or capacity to perform manually large-scale tasks such as scouting. The AI of the system can facilitate the processing of raw data from drones into clear information that enables the users take rapid decisions. The end-users of the proposed platform do not need to know all details of the infrastructure used to acquire data and control the physical world processes, nor the low level acquired data (sensor readings). Instead, they need processed information that closely represents physical and natural processes, and tools to predict the evolution of those processes given a concrete set of inputs. In this regard, the platform builds on the concept of Digital Twin, providing a logical separation between the low-level technical and use case-specific views of the information. On the one hand, the data acquired by sensors in the field, either using aerial embedded cameras, satellites, or any other means, are entered, tagged, stored, and classified by

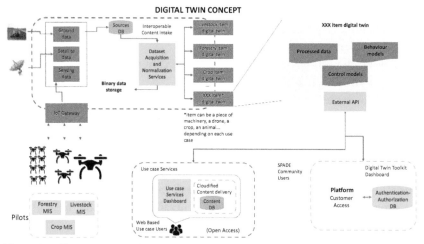

Figure 7.2 The concept and structure of the Digital Twin-based ecosystem.

the platform. On the other hand, the end-users are provided with an interface to perform high level actions on simulated and real-world assets, while they can access the information about those same assets that is relevant for their needs, regardless of the technical means used to acquire it. Finally, the platform also provides a means to abstract complex data acquisition (i.e., drone missions) through a mechanism that allows to orchestrate drone operations based on the information requested, including both fully automated and human-assisted drone missions.

In conclusion, one of the key elements of digital tools to manage agricultural and forestry environments is the development of a Digital Twin platform. This platform is designed around open-source technologies that will completely switch the paradigm on how information is obtained and used. Instead of focusing on the technical aspects of data acquisition, the platform provides an intent-based interface for end-users to require use case-specific information, and the process to register, transform, and provide that information is orchestrated by the platform in a transparent manner, therefore improving the user-centric nature of the technology. The structure of the digital twin platform and the ecosystem is provided in Fig. 7.2.

Discussion and main conclusions

Recent developments in unmanned aerial systems have allowed centimeter-level accuracy to be achieved without ground-based surveys, leading to pioneering applications in diverse fields. Among these applications, livestock farming, open-field agriculture, and forestry are some of the most promising ones, enabling efficient and sustainable management. However, these fields are characterized by a number of environmental, socio-economic, and technological challenges. UAVs, through exploiting advances in AI, ML, IoT, and other kinds of ICT, seem to be able to address these challenges and contribute toward the widespread adoption of drones. In this chapter, a brief discussion on the challenges, opportunities, and prospects associated with drones' use in the aforementioned sectors was first presented. Subsequently, a relevant concept for developing a holistic system for multipurpose physical-cyber agri-forest drones governance and environmental observation was described. This concept aims at developing a state-of-the-art ecosystem to tackle the multipurpose needs via leveraging UAVs to support sustainable digital services. These services refer to a wide range of different end-users in the sectors of livestock farming, open-field agriculture, and forestry. Finally, although the expected benefits of the adoption of drones cannot be argued, some crucial issues have to be addressed. These issues range from social acceptance, trust, data protection, privacy, regulation, safety, and security, as the governance of UAVs is a complicated area of legal regulation (Pagallo and Bassi, 2020; Nelson and Gorichanaz, 2019).

Acknowledgment

This work has been supported by the SPADE project, funded by the European Union's Horizon Europe Research and Innovation programme within HORIZON-CL6-2021-GOVERNANCE-01 under Grant Agreement no. 101060778.

References

Abbasi, R., Martinez, P., Ahmad, R., 2022. The digitization of agricultural industry — a systematic literature review on agriculture 4.0. Smart Agric. Technol. 2, 100042. Available from: https://doi.org/10.1016/j.atech.2022.100042.

Agrimonti, C., Lauro, M., Visioli, G., 2021. Smart agriculture for food quality: facing climate change in the 21st century. Crit. Rev. Food Sci. Nutr. 61, 971—981. Available from: https://doi.org/10.1080/10408398.2020.1749555.

Agriland Team, 2019. Checking and moving stock on the farm . . . with drones that bark. Available from: https://www.agriland.co.uk/farming-news/checking-and-moving-stock-on-the-farmwith-drones-that-bark/. (Accessed 6 Sept 2022).

Alanezi, M.A., Shahriar, M.S., Hasan, M.B., Ahmed, S., Sha'aban, Y.A., Bouchekara, H.R.E.H., 2022. Livestock management with unmanned aerial vehicles: a review. IEEE Access 10, 45001—45028. Available from: https://doi.org/10.1109/ACCESS.2022.3168295.

Alobid, M., Abujudeh, S., Szűcs, I., 2022. The role of blockchain in revolutionizing the agricultural sector. Sustainability 14. Available from: https://doi.org/10.3390/su14074313.

Anagnostis, A., Tagarakis, A.C., Kateris, D., Moysiadis, V., Sørensen, C.G., Pearson, S., et al., 2021. Orchard mapping with deep learning semantic segmentation. Sensors 21. Available from: https://doi.org/10.3390/S21113813.

Angelopoulou, T., Tziolas, N., Balafoutis, A., Zalidis, G., Bochtis, D., 2019. Remote sensing techniques for soil organic carbon estimation: a review. Remote Sens. 11. Available from: https://doi.org/10.3390/rs11060676.

Aquilani, C., Confessore, A., Bozzi, R., Sirtori, F., Pugliese, C., 2022. Review: precision livestock farming technologies in pasture-based livestock systems. Animal 16, 100429. Available from: https://doi.org/10.1016/j.animal.2021.100429.

Aslan, M.F., Durdu, A., Sabanci, K., Ropelewska, E., Gültekin, S.S., 2022. A comprehensive survey of the recent studies with UAV for precision agriculture in open fields and greenhouses. Appl. Sci. 12. Available from: https://doi.org/10.3390/app12031047.

Benalaya, N., Adjih, C., Laouiti, A., Amdouni, I., Saidane, L., 2022. UAV search path planning for livestock monitoring. In: Proceedings of the 2022 IEEE 11th IFIP International Conference on Performance Evaluation and Modeling in Wireless and Wired Networks (PEMWN); 1−6.

Benos, L., Bechar, A., Bochtis, D., 2020a. Safety and ergonomics in human-robot interactive agricultural operations. Biosyst. Eng. 200, 55−72. Available from: https://doi.org/10.1016/j.biosystemseng.2020.09.009.

Benos, L., Tsaopoulos, D., Bochtis, D., 2020b. A review on ergonomics in agriculture. part I: manual operations. Appl. Sci. 10, 1−21.

Benos, L., Bochtis, D.D., 2021. An analysis of safety and health issues in agriculture towards work automation. In: Bochtis, D.D., Pearson, S., Lampridi, M., Marinoudi, V., Pardalos, P.M. (Eds.), Information and Communication Technologies for Agriculture—Theme IV: Actions. Springer International Publishing, Cham, pp. 95−117. ISBN 978-3-030-84156-0.

Benos, L., Makaritis, N., Kolorizos, V., 2022a. From precision agriculture to agriculture 4.0: integrating ICT in farming. In: Bochtis, D.D., Sørensen, C.G., Fountas, S., Moysiadis, V., Pardalos, P.M. (Eds.), Information and Communication Technologies for Agriculture—Theme III: Decision. Springer International Publishing, Cham, pp. 79−93. ISBN 978-3-030-84152-2.

Benos, L., Sørensen, C.G., Bochtis, D., 2022b. Field deployment of robotic systems for agriculture in light of key safety, labor, ethics and legislation issues. Curr. Robot. Rep. Available from: https://doi.org/10.1007/s43154-022-00074-9.

Benos, L., Tagarakis, A.C., Dolias, G., Berruto, R., Kateris, D., Bochtis, D., 2021. Machine learning in agriculture: a comprehensive updated review. Sensors 21. Available from: https://doi.org/10.3390/S21113758.

Crist, E., Ripple, W.J., Ehrlich, P.R., Rees, W.E., Wolf, C., 2022. Scientists' warning on population. Sci. Total Environ. 845, 157166. Available from: https://doi.org/10.1016/j.scitotenv.2022.157166.

Cromwell, C., Giampaolo, J., Hupy, J., Miller, Z., Chandrasekaran, A., 2021. A systematic review of best practices for UAS data collection in forestry-related applications. Forests 12. Available from: https://doi.org/10.3390/f12070957.

Dash, J.P., Pearse, G.D., Watt, M.S., 2018. UAV multispectral imagery can complement satellite data for monitoring forest health. Remote Sens. 10. Available from: https://doi.org/10.3390/rs10081216.

Dian Bah, M., Hafiane, A., Canals, R., 2018. Deep learning with unsupervised data labeling for weed detection in line crops in UAV images. Remote Sens. Available from: https://doi.org/10.3390/rs10111690.

Dolias, G., Benos, L., Bochtis, D., 2022. On the routing of unmanned aerial vehicles (UAVs) in precision farming sampling missions. In: Bochtis, D.D., Sørensen, C.G., Fountas, S., Moysiadis, V., Pardalos, P.M. (Eds.), Information and Communication Technologies for Agriculture—Theme III: Decision. Springer International Publishing, Cham, pp. 95—124. ISBN 978-3-030-84152-2.

European Commission, 2022. The State of the World's Forests (SOFO).

Feduck, C., McDermid, G.J., Castilla, G., 2018. Detection of coniferous seedlings in UAV imagery. Forests 9. Available from: https://doi.org/10.3390/f9070432.

Fraser, B.T., Congalton, R.G., 2019. Evaluating the effectiveness of unmanned aerial systems (UAS) for collecting thematic map accuracy assessment reference data in New England forests. Forests 10. Available from: https://doi.org/10.3390/f10010024.

Gemtos, T., Fountas, S., Tagarakis, A., Liakos, V., 2013. Precision agriculture application in fruit crops: experience in handpicked fruits. Procedia Technol. Available from: https://doi.org/10.1016/j.protcy.2013.11.043.

Hasegawa, T., Sakurai, G., Fujimori, S., Takahashi, K., Hijioka, Y., Masui, T., 2021. Extreme climate events increase risk of global food insecurity and adaptation needs. Nat. Food 2, 587—595. Available from: https://doi.org/10.1038/s43016-021-00335-4.

Hassanalian, M., Abdelkefi, A., 2017. Classifications, applications, and design challenges of drones: a review. Prog. Aerosp. Sci. 91, 99—131. Available from: https://doi.org/10.1016/j.paerosci.2017.04.003.

Jhariya, M.K., Meena, R.S., Banerjee, A., 2021. Ecological intensification of natural resources towards sustainable productive system. In: Jhariya, M.K., Meena, R.S., Banerjee, A. (Eds.), Ecological Intensification of Natural Resources for Sustainable Agriculture. Springer Singapore, Singapore, pp. 1—28. ISBN 978-981-33-4203-3.

Ju, C., Son, H.I.L., 2018. Multiple UAV systems for agricultural applications: control, implementation, and evaluation. Electron 7, 1—19. Available from: https://doi.org/10.3390/electronics7090162.

Kalyani, Y., Collier, R., 2021. A systematic survey on the role of cloud, fog, and edge computing combination in smart agriculture. Sensors 21. Available from: https://doi.org/10.3390/s21175922.

Katikaridis, D., Moysiadis, V., Tsolakis, N., Busato, P., Kateris, D., Pearson, S., et al., 2022. UAV-supported route planning for UGVs in semi-deterministic agricultural environments. Agron. 12, 1937. Available from: https://doi.org/10.3390/AGRONOMY12081937.

Khaliq, A., Comba, L., Biglia, A., Ricauda Aimonino, D., Chiaberge, M., Gay, P., 2019. Comparison of satellite and UAV-based multispectral imagery for vineyard variability assessment. Remote Sens. 11. Available from: https://doi.org/10.3390/rs11040436.

Kim, J., Kim, S., Ju, C., Son, H.I.L., 2019. Unmanned aerial vehicles in agriculture: a review of perspective of platform, control, and applications. IEEE Access 7, 105100—105115. Available from: https://doi.org/10.1109/ACCESS.2019.2932119.

Labib, N.S., Brust, M.R., Danoy, G., Bouvry, P., 2021. The rise of drones in internet of things: a survey on the evolution, prospects and challenges of unmanned aerial vehicles. IEEE Access 9, 115466—115487. Available from: https://doi.org/10.1109/ACCESS.2021.3104963.

Lampridi, M., Sørensen, C., Bochtis, D., 2019a. Agricultural sustainability: a review of concepts and methods. Sustainability 11, 5120. Available from: https://doi.org/10.3390/su11185120.

Lampridi, M.G., Kateris, D., Vasileiadis, G., Marinoudi, V., Pearson, S., Sørensen, C.G., et al., 2019b. A case-based economic assessment of robotics employment in precision arable farming. Agronomy 9, 175. Available from: https://doi.org/10.3390/agronomy9040175.

Lampridi, M., Kateris, D., Sørensen, C.G., Bochtis, D., 2020. Energy footprint of mechanized agricultural operations. Energies 13. Available from: https://doi.org/10.3390/en13030769.

Liakos, K., Busato, P., Moshou, D., Pearson, S., Bochtis, D., Liakos, K.G., et al., 2018. Machine learning in agriculture: a review. Sensors 18, 2674. Available from: https://doi.org/10.3390/s18082674.

Liu, Y., Ma, X., Shu, L., Hancke, G.P., Abu-Mahfouz, A.M., 2021. From industry 4.0 to agriculture 4.0: current status, enabling technologies, and research challenges. IEEE Trans. Ind. Informatics 17, 4322−4334. Available from: https://doi.org/10.1109/TII.2020.3003910.

Maghazei, O., Netland, T., 2020. Drones in manufacturing: exploring opportunities for research and practice. J. Manuf. Technol. Manag. 31, 1237−1259. Available from: https://doi.org/10.1108/JMTM-03-2019-0099.

Marinoudi, V., Lampridi, M., Kateris, D., Pearson, S., Sørensen, C.G., Bochtis, D., 2021. The future of agricultural jobs in view of robotization. Sustainability 13. Available from: https://doi.org/10.3390/su132112109.

Matese, A., 2020. Editorial for the special issue "Forestry Applications of Unmanned Aerial Vehicles (UAVs)." Forests 11. Available from: https://doi.org/10.3390/f11040406.

Mazinani, M., Zarafshan, P., Dehghani, M., Vahdati, K., Etezadi, H., 2023. Design and analysis of an aerial pollination system for walnut trees. Biosyst. Eng. 225, 83−98. Available from: https://doi.org/10.1016/j.biosystemseng.2022.12.001.

Moysiadis, V., Tsolakis, N., Katikaridis, D., Sørensen, C.G., Pearson, S., Bochtis, D., 2020. Mobile robotics in agricultural operations: a narrative review on planning aspects. Appl. Sci. 10, 3453. Available from: https://doi.org/10.3390/app10103453.

Nelson, J., Gorichanaz, T., 2019. Trust as an ethical value in emerging technology governance: the case of drone regulation. Technol. Soc. 59, 101131. Available from: https://doi.org/10.1016/j.techsoc.2019.04.007.

Osinga, S.A., Paudel, D., Mouzakitis, S.A., Athanasiadis, I.N., 2022. Big data in agriculture: between opportunity and solution. Agric. Syst. 195, 103298. Available from: https://doi.org/10.1016/j.agsy.2021.103298.

Padró, J.-C., Muñoz, F.-J., Planas, J., Pons, X., 2019. Comparison of four UAV georeferencing methods for environmental monitoring purposes focusing on the combined use with airborne and satellite remote sensing platforms. Int. J. Appl. Earth Observ. Geoinf. 75, 130−140. Available from: https://doi.org/10.1016/j.jag.2018.10.018.

Pagallo, U., Bassi, E., 2020. The governance of unmanned aircraft systems (UAS): aviation law, human rights, and the free movement of data in the EU. Minds Mach. 30, 439−455. Available from: https://doi.org/10.1007/s11023-020-09541-8.

Popescu, D., Stoican, F., Stamatescu, G., Ichim, L., Dragana, C., 2020. Advanced UAV−WSN system for intelligent monitoring in precision agriculture. Sensors 20. Available from: https://doi.org/10.3390/s20030817.

Qiu, Z., Feng, Z.-K., Wang, M., Li, Z., Lu, C., 2018. Application of UAV photogrammetric system for monitoring ancient tree communities in Beijing. Forests 9. Available from: https://doi.org/10.3390/f9120735.

Radoglou-Grammatikis, P., Sarigiannidis, P., Lagkas, T., Moscholios, I., 2020. A compilation of UAV applications for precision agriculture. Comput. Networks 172, 107148. Available from: https://doi.org/10.1016/j.comnet.2020.107148.

Rejeb, A., Abdollahi, A., Rejeb, K., Treiblmaier, H., 2022. Drones in agriculture: a review and bibliometric analysis. Comput. Electron. Agric. 198, 107017. Available from: https://doi.org/10.1016/j.compag.2022.107017.

Schillings, J., Bennett, R., Rose, D.C., 2021. Exploring the potential of precision livestock farming technologies to help address farm animal welfare. Front. Anim. Sci. 13. Available from: https://doi.org/10.3389/fanim.2021.639678.

Sekhri, S., 2022. Agricultural trade and depletion of groundwater. J. Dev. Econ. 156, 102800. Available from: https://doi.org/10.1016/j.jdeveco.2021.102800.

Shin, J., Seo, W., Kim, T., Park, J., Woo, C., 2019. Using UAV multispectral images for classification of forest burn severity—a case study of the 2019 Gangneung forest fire. Forests 10. Available from: https://doi.org/10.3390/f10111025.

Singh, A.P., Yerudkar, A., Mariani, V., Iannelli, L., Glielmo, L., 2022. A bibliometric review of the use of unmanned aerial vehicles in precision agriculture and precision viticulture for sensing applications. Remote Sens. 14. Available from: https://doi.org/10.3390/rs14071604.

Sørensen, C.G., Fountas, S., Nash, E., Pesonen, L., Bochtis, D., Pedersen, S.M., et al., 2010. Conceptual model of a future farm management information system. Comput. Electron. Agric. 72, 37—47. Available from: https://doi.org/10.1016/j.compag.2010.02.003.

Sørensen, C.A.G., Kateris, D., Bochtis, D., 2019 ICT innovations and smart farming. In: Proceedings of the Communications in Computer and Information Science.

Tagarakis, A.C., Benos, L., Kateris, D., Tsotsolas, N., Bochtis, D., 2021. Bridging the gaps in traceability systems for fresh produce supply chains: overview and development of an integrated IoT-based system. Appl. Sci. 11. Available from: https://doi.org/10.3390/app11167596.

Tagarakis, A.C., Filippou, E., Kalaitzidis, D., Benos, L., Busato, P., Bochtis, D., 2022. Proposing UGV and UAV systems for 3D mapping of orchard environments. Sensors 22. Available from: https://doi.org/10.3390/s22041571.

Torresan, C., Berton, A., Carotenuto, F., Di Gennaro, S.F., Gioli, B., Matese, A., et al., 2017. Forestry applications of UAVs in Europe: a review. Int. J. Remote Sens. 38, 2427—2447. Available from: https://doi.org/10.1080/01431161.2016.1252477.

Tsouros, D.C., Bibi, S., Sarigiannidis, P.G., 2019. A review on UAV-based applications for precision agriculture. Information 10, 349. Available from: https://doi.org/10.3390/info10110349.

Vrchota, J., Pech, M., Švepešová, I., 2022. Precision agriculture technologies for crop and livestock production in the Czech Republic. Agriculture 12. Available from: https://doi.org/10.3390/agriculture12081080.

Wang, Y., Bai, R., Lu, X., Quan, S., Liu, Y., Lin, C., et al., 2022. Pollination parameter optimization and field verification of UAV-based pollination of 'Kuerle Xiangli'. Agronomy 12. Available from: https://doi.org/10.3390/agronomy12102561.

Zhai, Z., Martínez, J.F., Beltran, V., Martínez, N.L., 2020. Decision support systems for agriculture 4.0: survey and challenges. Comput. Electron. Agric. 170, 105256.

Zhao, L., Shi, Y., Liu, B., Hovis, C., Duan, Y., Shi, Z., 2019. Finer classification of crops by fusing UAV images and Sentinel-2A data. Remote Sens. 11. Available from: https://doi.org/10.3390/rs11243012.

SECTION D

Sustainability aspects

Information management infrastructures for multipurpose unmanned aerial systems operations

Lefteris Benos[1], Aristotelis C. Tagarakis[1], G. Vasileiadis[2,3], Dimitrios Kateris[1] and Dionysis Bochtis[1]

[1]Institute for Bio-Economy and Agri-Technology (IBO), Centre for Research and Technology Hellas (CERTH), Thessaloniki, Greece
[2]farmB Digital Agriculture S.A., Thessaloniki, Greece
[3]Department of Hydraulics, Soil Science and Agricultural Engineering, School of Agriculture, Aristotle University of Thessaloniki, Thessaloniki, Greece

Introduction

Nowadays, agriculture is weighed down by mitigating problems related to demographic, environmental, and socio-economic factors (Benos et al., 2022a). Besides, the unexpected negative impact posed by the recent COVID-19 pandemic (Okolie and Ogundeji, 2022; Bochtis et al., 2020) as well as wars and conflicts (Glauben et al., 2022; Pörtner et al., 2022; Ben Hassen and El Bilali, 2022) are jeopardizing both directly and indirectly global food security. As a consequence, there exists a pressing need to increase the efficiency of agricultural practices while reducing the environmental burden. These two conditions are mainly driving the evolution of agriculture toward establishing sustainability and a safe environment (Lampridi et al., 2019; Rose et al., 2021). In fact, "Agriculture 4.0" embodies the evolution of "precision agriculture" by addressing all actions on the basis of accurate analysis of data and information gathered and transmitted via advanced technologies (Debauche et al., 2022; Mühl and de Oliveira, 2022). Agricultural practices are no longer reliant on the uniform application of pesticides, fertilizers, and water to entire fields. On the contrary, farmers should make use of the minimum required quantities taking into account the fields' spatial variability and target the inputs to specific areas accordingly. Technological innovations have been introduced

Unmanned Aerial Systems in Agriculture
DOI: https://doi.org/10.1016/B978-0-323-91940-1.00008-6

gradually over the past decades, whereas the adoption of internet of things (IoT), has increased exponentially, accelerating the consolidation of Agriculture 4.0. This technology allows various tools, such as drones, sensors, cameras, robots, satellites, and agricultural machinery, to connect and communicate with each other sharing information and data that could be used for management purposes (Benos et al., 2021; Bochtis et al., 2014). The IoT and network environment is progressively shifting from wired networks to wireless communication. Meanwhile, the users' demand for IoT applications is changing to real-time and context-aware service delivery, resulting in a gradual shift in focus from the cloud to the edge (Ren et al., 2017; Tagarakis et al., 2021).

As a means to provide a digital place, where the numerous inputs from workplaces and IoT data are collected and fused in order to manage different aspects of agricultural production, the Digital Twins concept has emerged. The more data streams, the more information about the workplace. Without the Digital Twins, the IoT would require much more networking between the points of origin of the data and the points of use of the data. The terminology "Digital Twin" was first introduced by Grieves (Grieves and Vickers, 2017), and since then, the concept has witnessed an increase in popularity in several sectors, including agriculture. As a matter of fact, Digital Twin is acknowledged as a key driver of transition toward Industry 4.0 (Gürdür Broo et al., 2022; Aheleroff et al., 2021). A Digital Twin can be leveraged for the optimization of proposed and operational changes in a "secure" virtual environment, with no risk of unexpected and possibly detrimental effects on production. The simulated changes can subsequently be virtually replicated in the physical world if they are shown to be effective.

This chapter focuses on the application of Digital Twin in agriculture as a means of exploiting the diverse information coming from a variety of sensors integrated in unmanned aerial systems (UAS). To illustrate this topic, the remainder of the present chapter is structured as follows: The next section outlines the agricultural applications of UAS. A section follows that briefly describes the concept of Digital Twin from an agricultural perspective, highlighting also some recent applications. Subsequently, the available open-source platforms for Digital Twin are summarized that can be potentially used for hosting and processing of data provided by IoT sources, including UAS. Finally, concluding remarks are drawn in the last section along with a discussion from a broader viewpoint.

Agricultural applications of unmanned aerial systems

The deployment of UAS in the agricultural sector is part of an emerging tendency to integrate advanced technologies into farming activities that enable farmers to optimize their workflow and use resources more efficiently. In particular, key factors for the rapid evolution and usability of UAS in agricultural systems are the advances in the available sensor technology, the simplification and standardization of the operation procedures, combined with the decreasing prices and the improved flight durations due to increased battery performance. The UAV-enabled sensors provide outstanding observation capabilities across a wide range of agricultural operations, rendering UAS an alternative to the constraints of ground-based IoT infrastructure and satellite imagery (Tagarakis et al., 2022). Unlike those technologies, images taken by UAS generally display higher resolution in both temporal and spatial scales, and are unaffected by cloud cover, thus, making them suitable for precision farming operations (Zhang and Kovacs, 2012). Generally, sensors used for drones can detect energy at certain wavelengths of the electromagnetic spectrum and are categorized into two categories, namely passive and active. The former is the most commonly utilized type in agriculture and measure the energy reflected by the surface of interest. They are usually lightweight and relatively low cost. Nevertheless, they are affected by environmental conditions. In contrast, active sensors emit energy and then sense the reflection of that energy. Unlike passive sensors, they are not influenced by ambient conditions. However, they are heavier and more expensive (Erdle et al., 2011).

The term "agriculture" encompasses various production sectors ranging from crop and livestock to forestry, fisheries, and aquaculture. Focusing on the first three subsectors, UAS have found profile ground in several applications. Firstly, UAS are an ideal tool for monitoring and scouting, as they can cover large areas in relatively short time providing the ability to monitor crop and soil health and growth. Weed detection and mapping is a common application, since aerial remote sensing offers a number of advantages owing to its high flexibility in spatial analysis (van der Merwe et al., 2020). Aerial mapping can also support the identification and quantification of crop damage after weather events or other reasons, such as pests and diseases (Puri et al., 2017). Furthermore, UAS can provide accurate robust information, under various ambient conditions, reflecting water shortage, as well as soil properties (Angelopoulou et al., 2019). Specific types of UAS can also spray insecticides enabling site-specific applications, increasing the application efficiency, and minimizing the environmental impact. Another use of UASs is the application of fertilizers, a particularly useful feature for in-season site-specific corrective actions, or for the cases of difficult to reach areas. During these procedures, operators have full control of the UAS.

As far as livestock farming is concerned, aerial inspection of animals can contribute to define and isolate possibly sick animals (Benalaya et al., 2022; Vrchota et al., 2022). This can be accomplished through identifying animals that show symptoms of disease or measuring information like physical activity, weight, size, and color in an effective and timely manner compared to manual inspection. Moreover, UAS can help toward counting, monitoring, and tracking livestock animals as well as securing them against potential threats (Alanezi et al., 2022). Finally, concerning the forestry application, foresters were one of the first to introduce aerial remote sensing to monitor and map forest land, surveil illegal activities, monitor fires, and conserve biodiversity and wildlife (Cromwell et al., 2021; Guimarães et al., 2020).

Digital Twin paradigm in agriculture
General context of Digital Twin in agriculture

Generally speaking, Digital Twin refers to the capability to clone a physical entity into an equivalent environment (Minerva et al., 2020). In more detail, as stated by Haag and Anderl (2018): *"The Digital Twin is a comprehensive digital representation of an individual product and includes the properties, condition and behavior of the real-life object through models and data."* It comprises a set of realistic models, which are able to emulate the actual behavior. Digital Twin is developed in parallel with the physical twin and shall remain its virtual copy throughout its life cycle. Only in recent years, with the development of digital transformation and the advancement of tools, such as IoT, cloud, and machine learning, has this technology started to gain traction in practical use cases (Minerva et al., 2020; Qi and Tao, 2018; Tao et al., 2018; Liakos et al., 2018). Within the Industry 4.0, the concept of Digital Twin makes it possible to study and solve problems before they occur, prevent downtime, develop new opportunities, and even plan through simulations consisting of virtual tests that do not involve costs or loss of time (Gürdür Broo et al., 2022; Aheleroff et al., 2021). In short, the application of the Digital Twin technology creates an enormous disruptive milestone in the productivity of all types of

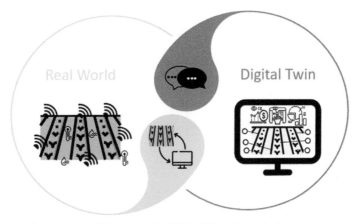

Figure 8.1 The conceptual framework of Digital Twin in agriculture.

organizations, becoming a fundamental tool to develop more efficient and more productive entities.

As can be depicted in Fig. 8.1, a Digital Twin framework is made up of three distinct elements:

1. The physical object (or process) and its surrounding environment;
2. Their digital representation; and
3. Their communication channel.

The interconnections between the physical and digital world involve information and data flows that integrate sensor flows between the physical and virtual environments.

Overall, Digital Twins need to be:

- Data informed;
- Near real time;
- Realistic; and
- Actionable.

Although the cost of developing and deploying Digital Twins is, in the first place, considerable, the expected benefits they may confer are manifold (Purcell and Neubauer, 2023; Neethirajan and Kemp, 2021; Singh et al., 2022; Rasheed et al., 2020). Digital Twins have the potential to give farmers and foresters the ability to address unforeseen discrepancies by identifying problems in advance, therefore, planning preventive measures in a timely and effective manner toward providing immediate solutions to mitigate problems. The implementation benefits are listed in Fig. 8.2.

Figure 8.2 Implementation benefits of introducing Digital Twin in agriculture.

A key feature of Digital Twin is the fusion of information. In particular, Digital Twins integrate information from diverse and heterogeneous sources. In addition, they examine physical twins from different angles by using numerous data sources and evaluating possible action responses. Lastly, Digital Twins may show human-centered intelligence in control mechanisms for previously neglected aspects, such as human–machine interaction for more secure work environments (Pylianidis et al., 2021; Benos et al., 2020; Anagnostis et al., 2021).

An overview of cyber-physical systems based on Digital Twins for enabling smart agriculture

Despite the growing interest of implementing Digital Twins in agriculture, only a few applications are available in the relative literature. The literature ranges from studies focusing on the agri-food supply chains (Defraeye et al., 2021; Tebaldi et al., 2021; van der Valk et al., 2022), open-field crop production (Moghadam et al., 2019; Cobbenhagen et al., 2021; Verdouw and Kruize, 2017), controlled agricultural environment (Slob and Hurst, 2022; Chaux et al., 2021; Ariesen-Verschuur et al., 2022), livestock farming (Neethirajan and Kemp, 2021; Jo et al., 2019,

2018; Han et al., 2022), and forestry (Niță, 2021; Buonocore et al., 2022), to studies aiming at investigating societal and ethical raising issues (van der Burg et al., 2021).

One indicative use case deploying a Digital Twin framework is that of Jo et al. (2018), where a design concept was presented pertaining to a pig farm that was the physical twin. The aim was to monitor the health status of animals and prevent diseases. The following process takes place in Jo et al. (2018): (1) the sensed data collected in the physical system (NH_3, CO_2, temperature, dust, humidity, etc.) are introduced in the Digital Twin (Big Data component); (2) the simulations (using the aforementioned stored data and deep learning methods) running on the Digital Twin suggest optimal conditions of the livestock farm (e.g., optimal temperature and CO_2); (3) then a control system applies targeted actions to the physical system toward accomplishing the aforementioned optimal conditions (for this purpose, open windows and fans are used during certain periods); and (4) the results are presented in a properly designed intuitive interface.

Another demonstrative study, this time concerning orchards, is that of Moghadam et al. (2019), where an automated monitoring system, namely AgScan3D + , was developed in order to create a Digital Twin of every tree. This system is composed of a rotating 3D LiDAR (three-dimensional light detection and ranging) and cameras, which can be mounted on an agricultural vehicle toward providing real-time decision support, through monitoring the status of each plant, including structure, health, stress, and fruit quality. AgScan3D + was successfully tested in macadamia, mango, and avocado orchards and modeled useful canopy features, like light penetration distribution and foliage density.

As regards forest management through Digital Twins, an indicative study was presented in Niță (2021). An automatic workflow was developed and validated regarding the processing of 3D point clouds so as to provide Digital Twins for each tree with a GeoSLAM mobile LiDAR scanner along with an artificial intelligence-based platform.

Digital Twins for leveraging the utilization of unmanned aerial systems in agriculture

Concerning the utilization of UAS in agriculture by means of Digital Twins, the literature is very scarce, mainly at a concept level, and does not cover large-scale applications. In Teschner et al. (2022), for instance, a drone-based intrusion detection workflow was described. More specifically, a Digital Twin was developed for emulating the UAS and its surrounding environment with the intention of protecting agricultural fields from wild animal entering.

In another study, Tsolakis et al. (2021) developed an approach for facilitating the use of UAS to monitor the water stress status of individual trees and collect information related to irrigation activities. For that purpose, a 3D conceptual orchard (Digital Twin) was created at the Gazebo simulation environment along with a quadrotor drone. The simulated UAS had the ability to rotate, hover, and detect freshwater needs using the robot operating system (ROS). The UAS could navigate within the orchard autonomously on account of a routing algorithm and the signals from the sensors of the cyber space (Dolias et al., 2022; Tsolakis et al., 2019). In addition to the emulated environment, a first approach was made to perform a real-world pilot testing through using band combinations from the multispectral camera mounted on a fixed-wing UAS. Furthermore, a rotary-wing drone was utilized to monitor the condition of the irrigation equipment.

The study of Tagarakis et al. (2022) used UAS for mapping the spatial properties and geometry of trees and objects in commercial orchards providing a 3D digital representation of the orchard's environment using photogrammetry. The outcomes were compared against those provided by an unmanned ground vehicle equipped with an RGB-D (red green blue-depth) camera. In total, the 3D orthomosaics acquired by the UAS resulted in similar tree height measurements with the ground vehicle, whereas the 3D point cloud created from the ground system proved to be more detailed. The authors concluded that the fusion of the two datasets could complement each other by providing an accurate tree representation. Moreover, the fused 3D point clouds can be a very useful input for a simulation environment, similarly to Tsolakis et al. (2021), to provide a Digital Twin toward demonstrating the real-time navigation and interaction of robots within the orchard environment.

The concept of incorporating Digital Twins in support of utilizing UAS in agriculture is quite new. It is the main framework of a European initiative, a research and innovation action project entitled "Multi-purpose physical-cyber agri-forest drones' ecosystem for governance and environmental observation - SPADE." The project aims at leveraging the potential advantages that can be derived from the use of UAS in agricultural applications. Thus, a digital platform is being developed to make UAS operations more accessible, offering a service channel for value-added

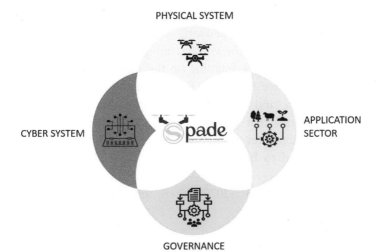

Figure 8.3 The main components of SPADE EU initiative including the physical system and the corresponding cyber one (Digital Twin) as well as awareness on governance models for exploring potential unmanned aerial systems applications in the sectors of crop production, livestock farming, and forestry.

services in crop production, livestock farming, and forestry sectors (Fig. 8.3). The platform is based on the concept of a Digital Twin, where the data acquired by sensors mounted on UAS and sensors in the field, or other sources, are entered, analyzed, and processed. The Digital Twin platform also provides the stakeholders with a user-friendly interface for executing high-level operations on emulated and real assets. Interestingly, the SPADE platform is being constructed around open-source technologies in a transparent and user-centric manner. To sum up, the system's artificial intelligence can enable the processing of raw data from UAS, among others, into meaningful information allowing users to make quick and reliable decisions without focusing on the technical aspects.

Open-source Digital Twins and internet of things platforms for facilitating the unmanned aerial systems deployment in agriculture

In order to address the necessity for economic development, software companies, such as Microsoft and Amazon, among others, have started to provide support for the set-up and operation of Digital Twin, typically referred to as Digital Twin platforms. However, open-source initiatives have also offered alternatives allowing for collaboration on the development of the Digital Twin and IoT concepts. Open source is a form of licensing agreement allowing users to freely modify a work, exploit that work in many new ways, integrate the work into a larger work, or create a new work building on the original work. By removing the barriers, open source fosters the free sharing of ideas in a community to facilitate scientific and technological progress. Table 8.1 summarizes open-source platforms that enable Digital Twins development and IoT integration and interaction along with related information. These platforms are anticipated to host, among others, data from UAS for several agricultural applications.

Based on Lehner et al. (2022), there are 13 requirements for the Digital Twin platforms:

- Synchronization;
- Convergence;
- Verification and validation;
- Real-time behavior;
- Automation protocols;
- Platform interoperability;
- System interoperability;
- Domain expert involvement;
- Connection and data security;
- Modifiability;
- Reusability;
- Continuous integration and deployment; and
- Provisioning.

Lehner et al. (2022) concluded that the existing capabilities of Digital Twin platforms meet most of the above requirements and, therefore, can offer many advantages to stakeholders. Furthermore, while implementation via open-source platforms was more efficient than developing the platform from scratch, some functions had to be manually developed, that could be generalizable, leaving room for further improvements in the future. Finally, proper user interfaces are needed for developing value-adding services through machine learning and simulation methods combined with their integration into Digital Twins systems.

Discussion and main conclusions

In the present chapter, the Digital Twin concept was presented through the lens of Agriculture 4.0. In particular, this study tried to shed light on

Table 8.1 List of open-source Digital Twins and internet of things (IoT) platforms.

Platform	Related information
Azure Digital Twins	An IoT platform that enables easy creation of a digital representation of real things, such as farms, to optimize processes and associated costs. It offers an open modeling language to create custom domain models and powerful application programming interfaces (APIs) and integrates with Azure data analytics. Developer: Microsoft
iModel.js	It was developed to be flexible and open for easier access, utilization, and integration of individual Digital Twins with other systems. It also enhances the visualization and comprehension of physical assets. Developer: Bentley
Eclipse Ditto	Ditto is a framework that supports the implementation of IoT Digital Twins software standards. The capabilities of Ditto: (1) mirrors physical assets/sensors, (2) acts as a "single source of truth" for a physical asset, (3) provides services around devices, and (4) keeps the physical and digital worlds synchronized. Its Digital Twin framework enables web-based APIs for interacting with Digital Twins, assures that access can only be granted by authorization, allows one-to-one or one-to-many Digital Twin interactions, and integrates with other back-end infrastructure such as brokers and messaging systems. Developer: Eclipse
Eclipse Hono	It enables the use of remote service interfaces to connect a wide variety of IoT devices and interact with them in a unified way, regardless of the device's communication protocol. It also supports devices that communicate via common IoT protocols. Hono is a microservices-based architecture that makes use of a reactive programming model, providing horizontal scalability. Therefore communication is simplified through an API, regardless of the device protocol. Developer: Eclipse
Eclipse CHE	It offers a centralized development environment running on Kubernetes or OpenShift and a multicontainer workspace for each developer with one-click replication using Eclipse CHE options. Prebuilt stacks allow for the creation of custom stacks for any language or runtime. It also supports tool protocols such as Language Server Protocol or Debug Adapter Protocol, a plugin engine compatible with Visual Studio Code extensions, and a software development kit (SDK) for creating custom cloud development platforms. Developer: Eclipse

(Continued)

Table 8.1 (Continued)

Platform	Related information
Eclipse THEIA	It is a scalable platform for developing integrated, multilanguage, cloud, and desktop IDE-like products with advanced web technologies. It is implemented in TypeScript, CSS, and HTML. THEIA consists of a front-end that runs in a browser or in the local desktop application and a back-end that runs on any host or locally within the desktop application. The front-end and back-end communicate via JSON RPC over websockets. A THEIA application consists of a number of extensions, which can contribute to both the front-end and back-end. Developer: Eclipse
Eclipse hawkBit	It is a domain-independent back-end framework for rolling out software updates to limited edge devices and more powerful controllers and gateways connected to IP-based networking infrastructures. Developer: Eclipse
Eclipse LESHAN	It is a Java implementation of the OMA Lightweight M2M server and client. It is not a stand-alone server, but rather a set of java libraries that aids people develop their own lightweight M2M server client. The project also includes a client, a server, and a bootstrap server demo as an example of its API and for testing purposes. Developer: Eclipse
Eclipse MOSQUITTO	Eclipse Mosquitto delivers a lightweight server implementation of the MQTT protocol that is appropriate for every situation; from full-power machines to embedded and low-power machines. Sensors and actuators, which are often the sources and destinations of MQTT messages, can be very small and lack power. This is also true for the embedded machines to which they are connected, where Mosquitto could be executed. Developer: Eclipse
Eclipse Kura	It is an open-source, scalable IoT edge framework based on Java/OSGi. It provides API access to the hardware interfaces of IoT gateways (serial ports, global positioning system (GPS), watchdog, general-purpose input/output (GPIO, inter-integrated circuit (I2C), etc.). It features ready-to-use field protocols (including Modbus, OPC-UA, S7), an application container and a web-based visual dataflow programming to acquire data from the field, process it at the edge, and publish it to leading IoT Cloud platforms via MQTT connectivity. Developer: Eclipse

(*Continued*)

Table 8.1 (Continued)

Platform	Related information
Apache Kafka	It is an open-source distributed event streaming platform used by many companies for high-performance data pipelines, data integration, flow analytics, and mission-critical applications. Kafka is a distributed system consisting of servers and clients communicating over a high-performance transmission control protocol (TCP) network protocol. It can be implemented on bare-metal hardware, virtual machines, and containers in on-premise and cloud environments. Developer: ASF
Influx DB	It is a real-time series database that is simple to set up and scale. A time series database is used to store logs, multiple types of data (e.g., sensor data) over a period. The basic function of the Influx DB is to record data in real time and read in real time, with a capacity of one million writers per second. Developer: influxdata
Grafana	Grafana is an analysis and monitoring solution that offers plug-in data source models and support for many of the most popular time series databases, including Graphite, Prometheus, Elastic search, and Influx DB. It also allows the following features: users can query, visualize, and alert on metrics and logs from multiple stored locations that are stored. Developer: Grafana Labs
Predix	Predix is the software platform for Digital Twin development, while it collects and analyzes data from industrial machines. It provides a cloud-based PaaS (platform as a Service) optimization and performance management. It also connects data, people, and equipment in a standardized way. Predix was originally designed to target factories and gave their ecosystems the same simple functionality as the operating systems that transformed mobile phones. Developer: General Electrics
OpenRemote	It can connect any device or data source for asset management to a single interface. It uses built-in tools to process data and create rules for asset maintenance. It can also convert data into information and create applications that fit users' workflow. It can also compare assets and evaluate their behavior over time with the data visualization dashboard and create reports to measure project progress. Developer: OpenRemote

(Continued)

Table 8.1 (Continued)

Platform	Related information
ThingsBoard	Features of ThingsBoard include: (1) IoT device management with monitoring and control mechanisms; (2) scalability with the ability to orchestrate multiple devices simultaneously, across the entire IoT ecosystem; (3) allowing users to create and manage alerts for connected devices (e.g., in case of disconnection or inactivity), other assets, and customers with real-time alarm monitoring; (4) extending the default functionality with customizable rule chains, widgets, and transport implementations; (5) multitenancy. Developer: ThingsBoard
Thinger.io	It is a scalable IoT cloud platform for connecting devices. You have an easy-to-navigate ready-to-use cloud infrastructure that allows users to integrate, monitor, and control millions of IoT devices. The platform is easy to use and easy to understand without having complex codes. The platform is hardware-agnostic. It can connect any device, the most typical being Arduino, ESP8266, Raspberry Pi, and Intel Edison. It requires only to install the server in the cloud and use the open-source libraries for integrating the IoT devices. Developer: Thinger.io
MainFlux	Mainflux is an open-source IoT platform that focuses on edge computing. Mainflux is patent-free and end-to-end, under an Apache 2.0 license, and covering most of the things needed for developing IoT solutions, applications, and products. It provides the complete infrastructure and middleware to execute: (1) device management; (2) data aggregation and data management; (3) connectivity and message routing; (4) IoT application enablement; (5) analytics. Developer: Mainflux Labs
Watson IoT	It creates multiple views of products by bringing together various digital threads and data streams that can be tailored to the needs of a particular user. Digital thread capabilities, combined with International Business Machines Corporation (IBM) engineering capabilities help individuals or teams understand the impact of engineering decisions on a products performance. Developer: IBM

applications of Digital Twin combined with IoT, particularly agri–forest UAS. To that end, after briefly discussing how data acquired from UAS are exploited in agricultural sector, the general context of Digital Twin was analyzed along with applications found in literature. Subsequently, a number of open-source Digital Twin and IoT platforms, which can potentially be used to integrate information from UAS, was presented.

It can be deduced that Digital Twins and their application in the agricultural sector are still in the early demonstration stages. Consequently, they cannot offer, currently, the benefits demonstrated in other fields. Remarkably, there is an important challenge when considering Digital Twins in agriculture. Unlike industrial cases, most agricultural processes involve living objects, including plants and animals or perishable products. Thus, developing Digital Twins for such kinds of systems is very demanding. Overall, there is a lack of references covering the large-scale application of this technology to heterogeneous crops with different input and output requirements. Moreover, the most Digital Twin paradigms in agriculture are at prototype or concept level (Pylianidis et al., 2021; Slob and Hurst, 2022; Nasirahmadi and Hensel, 2022). The reason behind this finding is not only the complex nature of agricultural systems, but also the fact that the agricultural sector itself is a slow adopter of technology (Delgado et al., 2019).

Most of the information and communication technologies (ICT) are still regarded as fields of experimentation in agriculture, despite the benefits that they can potentially offer to stakeholders (Basso and Antle, 2020). Furthermore, Digital Twin realization requires vast amounts of reliable data. However, at the moment, the sector is characterized by a lack of data culture (Jones et al., 2017), which inevitably decelerates the adoption of Digital Twin in agriculture. As a consequence, for a community which is less information technology oriented, this is a significant deterrent (Brown et al., 2019). On the other hand, IoT and other digital technologies are being increasingly adopted for numerous applications in agriculture. One of the main constraints in the further adoption of such systems, which also is one of the barriers in being included in Digital Twins, is the limited interoperability between the different devices. Interoperable solutions must be in place to allow secure and trusted external access to specific aspects of the Digital Twins (Benos et al., 2022b). External stakeholders can enrich them with a multitude of records, such as historical, soil, water, meteorological, and satellite data. In a nutshell, considering the benefits of adopting Digital Twins and IoTs in agriculture, they are anticipated to bring a technological revolution in the near future, provided that agricultural community is sufficiently informed and convinced to invest in them.

Acknowledgment

This work has been supported by the SPADE project, funded by the European Union's Horizon Europe Research and Innovation programme within HORIZON-CL6-2021-GOVERNANCE-01 under Grant Agreement no. 101060778.

References

Aheleroff, S., Xu, X., Zhong, R.Y., Lu, Y., 2021. Digital twin as a service (DTaaS) in industry 4.0: an architecture reference model. Adv. Eng. Inform. 47, 101225. Available from: https://doi.org/10.1016/j.aei.2020.101225.

Alanezi, M.A., Shahriar, M.S., Hasan, M.B., Ahmed, S., Sha'aban, Y.A., Bouchekara, H.R.E. H., 2022. Livestock management with unmanned aerial vehicles: a review. IEEE Access 10, 45001−45028. Available from: https://doi.org/10.1109/ACCESS.2022.3168295.

Anagnostis, A., Benos, L., Tsaopoulos, D., Tagarakis, A., Tsolakis, N., Bochtis, D., 2021. Human activity recognition through recurrent neural networks for human-robot interaction in agriculture. Appl. Sci. 11, 2188. Available from: https://doi.org/10.3390/app11052188.

Angelopoulou, T., Tziolas, N., Balafoutis, A., Zalidis, G., Bochtis, D., 2019. Remote sensing techniques for soil organic carbon estimation: a review. Remote Sens. 11, 676. Available from: https://doi.org/10.3390/rs11060676.

Ariesen-Verschuur, N., Verdouw, C., Tekinerdogan, B., 2022. Digital twins in greenhouse horticulture: a review. Comput. Electron. Agric. 199, 107183. Available from: https://doi.org/10.1016/j.compag.2022.107183.

Basso, B., Antle, J., 2020. Digital agriculture to design sustainable agricultural systems. Nat. Sustain. 3, 254−256. Available from: https://doi.org/10.1038/s41893-020-0510-0.

Benalaya, N., Adjih, C., Laouiti, A., Amdouni, I., Saidane, L., 2022 UAV search path planning for livestock monitoring. In: Proceedings of the 2022 IEEE 11th IFIP International Conference on Performance Evaluation and Modeling in Wireless and Wired Networks (PEMWN); 1−6.

Benos, L., Bechar, A., Bochtis, D., 2020. Safety and ergonomics in human-robot interactive agricultural operations. Biosyst. Eng. 200, 55−72. Available from: https://doi.org/10.1016/j.biosystemseng.2020.090.009.

Benos, L., Tagarakis, A.C., Dolias, G., Berruto, R., Kateris, D., Bochtis, D., 2021. Machine learning in agriculture: a comprehensive updated review. Sensors 21. Available from: https://doi.org/10.3390/S21113758.

Benos, L., Makaritis, N., Kolorizos, V., 2022a. From precision agriculture to agriculture 4.0: integrating ICT in farming. In: Bochtis, D.D., Sørensen, C.G., Fountas, S., Moysiadis, V., Pardalos, P.M. (Eds.), Information and Communication Technologies for Agriculture—Theme III: Decision. Springer International Publishing, Cham, pp. 79−93. ISBN 978-3-030-84152-2.

Benos, L., Sørensen, C.G., Bochtis, D., 2022b. Field deployment of robotic systems for agriculture in light of key safety, labor, ethics and legislation issues. Curr. Robot. Rep. Available from: https://doi.org/10.1007/s43154-022-00074-9.

Bochtis, D.D., Sørensen, C.G.C., Busato, P., 2014. Advances in agricultural machinery management: a review. Biosyst. Eng. 126, 69–81.

Bochtis, D., Benos, L., Lampridi, M., Marinoudi, V., Pearson, S., Sørensen, C.G., 2020. Agricultural workforce crisis in light of the COVID-19 pandemic. Sustainability 12. Available from: https://doi.org/10.3390/su12198212.

Brown, P., Daigneault, A., Dawson, J., 2019. Age, values, farming objectives, past management decisions, and future intentions in New Zealand agriculture. J. Environ. Manage. 231, 110–120. Available from: https://doi.org/10.1016/j.jenvman.2018.100.018.

Ben Hassen, T., El Bilali, H., 2022. Impacts of the Russia-Ukraine war on global food security: towards more sustainable and resilient food systems? Foods 11. Available from: https://doi.org/10.3390/foods11152301.

Buonocore, L., Yates, J., Valentini, R., 2022. A proposal for a forest digital twin framework and its perspectives. Forests 13. Available from: https://doi.org/10.3390/f13040498.

Chaux, J.D., Sanchez-Londono, D., Barbieri, G., 2021. A digital twin architecture to optimize productivity within controlled environment agriculture. Appl. Sci. 11.

Cobbenhagen, A.T.J.R., Antunes, D.J., van de Molengraft, M.J.G., Heemels, W.P.M.H., 2021. Opportunities for control engineering in arable precision agriculture. Annu. Rev. Control. 51, 47–55. Available from: https://doi.org/10.1016/j.arcontrol.2021.010.001.

Cromwell, C., Giampaolo, J., Hupy, J., Miller, Z., Chandrasekaran, A., 2021. A systematic review of best practices for UAS data collection in forestry-related applications. Forests 12. Available from: https://doi.org/10.3390/f12070957.

Debauche, O., Mahmoudi, S., Manneback, P., Lebeau, F., 2022. Cloud and distributed architectures for data management in agriculture 4.0: review and future trends. J. King Saud. Univ. Comput. Inf. Sci. 34, 7494–7514. Available from: https://doi.org/10.1016/j.jksuci.2021.090.015.

Defraeye, T., Shrivastava, C., Berry, T., Verboven, P., Onwude, D., Schudel, S., et al., 2021. Digital twins are coming: will we need them in supply chains of fresh horticultural produce? Trends Food Sci. Technol. 109, 245–258. Available from: https://doi.org/10.1016/j.tifs.2021.010.025.

Delgado, J.A., Short, N.M., Roberts, D.P., Vandenberg, B., 2019. Big data analysis for sustainable agriculture on a geospatial cloud framework. Front. Sustain. Food Syst. 3. Available from: https://doi.org/10.3389/fsufs.2019.00054.

Dolias, G., Benos, L., Bochtis, D., 2022. On the routing of unmanned aerial vehicles (UAVs) in precision farming sampling missions. In: Bochtis, D.D., Sørensen, C.G., Fountas, S., Moysiadis, V., Pardalos, P.M. (Eds.), Information and Communication Technologies for Agriculture—Theme III: Decision. Springer International Publishing, Cham, pp. 95–124. ISBN 978-3-030-84152-2.

Erdle, K., Mistele, B., Schmidhalter, U., 2011. Comparison of active and passive spectral sensors in discriminating biomass parameters and nitrogen status in wheat cultivars. F. Crop. Res. 124, 74–84. Available from: https://doi.org/10.1016/j.fcr.2011.060.007.

Glauben, T., Svanidze, M., Götz, L., Prehn, S., Jamali Jaghdani, T., Đurić, I., et al., 2022. The war in Ukraine, agricultural trade and risks to global food security. Intereconomics 57, 157–163. Available from: https://doi.org/10.1007/s10272-022-1052-7.

Grieves, M., Vickers, J., 2017. Digital twin: mitigating unpredictable, undesirable emergent behavior in complex systems. In: Kahlen, F.-J., Flumerfelt, S., Alves, A. (Eds.), Transdisciplinary Perspectives on Complex Systems: New Findings and Approaches. Springer International Publishing, Cham, pp. 85–113. ISBN 978-3-319-38756-7.

Guimarães, N., Pádua, L., Marques, P., Silva, N., Peres, E., Sousa, J.J., 2020. Forestry remote sensing from unmanned aerial vehicles: a review focusing on the data, processing and potentialities. Remote Sens. 12. Available from: https://doi.org/10.3390/rs12061046.

Gürdür Broo, D., Bravo-Haro, M., Schooling, J., 2022. Design and implementation of a smart infrastructure digital twin. Autom. Constr. 136, 104171. Available from: https://doi.org/10.1016/j.autcon.2022.104171.

Haag, S., Anderl, R., 2018. Digital twin — proof of concept. Manuf. Lett. 15, 64—66. Available from: https://doi.org/10.1016/j.mfglet.2018.020.006.

Han, X., Lin, Z., Clark, C., Vucetic, B., Lomax, S., 2022. AI based digital twin model for cattle caring. Sensors 22. Available from: https://doi.org/10.3390/s22197118.

Jones, J.W., Antle, J.M., Basso, B., Boote, K.J., Conant, R.T., Foster, I., et al., 2017. Toward a new generation of agricultural system data, models, and knowledge products: state of agricultural systems science. Agric. Syst. 155, 269—288. Available from: https://doi.org/10.1016/j.agsy.2016.090.021.

Jo, S.-K., Park, D.-H., Park, H., Kim, S.-H., 2018 Smart livestock farms using digital twin: feasibility study. In: Proceedings of the 2018 International Conference on Information and Communication Technology Convergence (ICTC); 1461—1463.

Jo, S.-K., Park, D.-H., Park, H., Kwak, Y., Kim, S.-H., 2019 Energy planning of pigsty using digital twin. In: Proceedings of the 2019 International Conference on Information and Communication Technology Convergence (ICTC); 723—725.

Lampridi, M., Sørensen, C., Bochtis, D., 2019. Agricultural sustainability: a review of concepts and methods. Sustainability 11, 5120. Available from: https://doi.org/10.3390/su11185120.

Lehner, D., Pfeiffer, J., Tinsel, E., Strljic, M., Sint, S., Vierhauser, M., et al., 2022. Digital twin platforms: requirements, capabilities, and future prospects. IEEE Softw. 39, 53—61. Available from: https://doi.org/10.1109/MS.2021.3133795.

Liakos, K., Busato, P., Moshou, D., Pearson, S., Bochtis, D., 2018. Machine learning in agriculture: a review. Sensors 18, 2674. Available from: https://doi.org/10.3390/s18082674.

Minerva, R., Lee, G.M., Crespi, N., 2020. Digital twin in the IoT context: a survey on technical features, scenarios, and architectural models. Proc. IEEE 108, 1785—1824. Available from: https://doi.org/10.1109/JPROC.2020.2998530.

Moghadam, P., Lowe, T., Edwards, E.J., 2019. Digital twin for the future of orchard production systems. Proceedings 36.

Mühl, D.D., de Oliveira, L., 2022. A bibliometric and thematic approach to agriculture 4.0. Heliyon 8, e09369. Available from: https://doi.org/10.1016/j.heliyon.2022.e09369.

Nasirahmadi, A., Hensel, O., 2022. Toward the next generation of digitalization in agriculture based on digital twin paradigm. Sensors 22, 498. Available from: https://doi.org/10.3390/s22020498.

Neethirajan, S., Kemp, B., 2021. Digital twins in livestock farming. Animals 11. Available from: https://doi.org/10.3390/ani11041008.

Niţă, M.D., 2021. Testing forestry digital twinning workflow based on mobile LiDAR scanner and AI platform. Forests 12. Available from: https://doi.org/10.3390/f12111576.

Okolie, C.C., Ogundeji, A.A., 2022. Effect of COVID-19 on agricultural production and food security: a scientometric analysis. Humanit. Soc. Sci. Commun. 9, 64. Available from: https://doi.org/10.1057/s41599-022-01080-0.

Purcell, W., Neubauer, T., 2023. Digital twins in agriculture: a state-of-the-art review. Smart Agric. Technol. 3, 100094. Available from: https://doi.org/10.1016/j.atech.2022.100094.

Puri, V., Nayyar, A., Raja, L., 2017. Agriculture drones: a modern breakthrough in precision agriculture. J. Stat. Manag. Syst. 20, 507—518. Available from: https://doi.org/10.1080/09720510.2017.1395171.

Pylianidis, C., Osinga, S., Athanasiadis, I.N., 2021. Introducing digital twins to agriculture. Comput. Electron. Agric. 184, 105942. Available from: https://doi.org/10.1016/j.compag.2020.105942.

Pörtner, L.M., Lambrecht, N., Springmann, M., Bodirsky, B.L., Gaupp, F., Freund, F., et al., 2022. We need a food system transformation—in the face of the Russia-Ukraine war, now more than ever. One Earth 5, 470−472. Available from: https://doi.org/10.1016/j.oneear.2022.040.004.

Qi, Q., Tao, F., 2018. Digital twin and big data towards smart manufacturing and industry 4.0: 360 degree comparison. IEEE Access 6, 3585−3593. Available from: https://doi.org/10.1109/ACCESS.2018.2793265.

Rasheed, A., San, O., Kvamsdal, T., 2020. Digital twin: values, challenges and enablers from a modeling perspective. IEEE Access 8, 21980−22012. Available from: https://doi.org/10.1109/ACCESS.2020.2970143.

Ren, J., Guo, H., Xu, C., Zhang, Y., 2017. Serving at the edge: a scalable IoT architecture based on transparent computing. IEEE Netw. 31, 96−105. Available from: https://doi.org/10.1109/MNET.2017.1700030.

Rose, D.C., Wheeler, R., Winter, M., Lobley, M., Chivers, C.A., 2021. Agriculture 4.0: making it work for people, production, and the planet. Land Use Policy 100, 104933. Available from: https://doi.org/10.1016/j.landusepol.2020.104933.

Singh, M., Srivastava, R., Fuenmayor, E., Kuts, V., Qiao, Y., Murray, N., et al., 2022. Applications of digital twin across industries: a review. Appl. Sci. 12. Available from: https://doi.org/10.3390/app12115727.

Slob, N., Hurst, W., 2022. Digital twins and industry 4.0 technologies for agricultural greenhouses. Smart Cities 5, 1179−1192. Available from: https://doi.org/10.3390/smartcities5030059.

Tagarakis, A.C., Benos, L., Kateris, D., Tsotsolas, N., Bochtis, D., 2021. Bridging the gaps in traceability systems for fresh produce supply chains: overview and development of an integrated IoT-based system. Appl. Sci. 11. Available from: https://doi.org/10.3390/app11167596.

Tagarakis, A.C., Filippou, E., Kalaitzidis, D., Benos, L., Busato, P., Bochtis, D., 2022. Proposing UGV and UAV systems for 3D mapping of orchard environments. Sensors 22. Available from: https://doi.org/10.3390/s22041571.

Tao, F., Cheng, J., Qi, Q., Zhang, M., Zhang, H., Sui, F., 2018. Digital twin-driven product design, manufacturing and service with big data. Int. J. Adv. Manuf. Technol. 94, 3563−3576. Available from: https://doi.org/10.1007/s00170-017-0233-1.

Tebaldi, L., Vignali, G., Bottani, E., 2021. Digital twin in the agri-food supply chain: a literature review. In: Dolgui, A., Bernard, A., Lemoine, D., von Cieminski, G., Romero, D. (Eds.), Advances in Production Management Systems. Artificial Intelligence for Sustainable and Resilient Production Systems. Springer International Publishing, Cham, pp. 276−283.

Teschner, G., Hajdu, C., Hollósi, J., Boros, N., Kovács, A., Ballagi, Á., 2022 Digital twin of drone-based protection of agricultural areas. In: Proceedings of the 2022 IEEE 1st International Conference on Internet of Digital Reality (IoD); 99−104.

Tsolakis, N., Bechtsis, D., Bochtis, D., 2019. Agros: a robot operating system based emulation tool for agricultural robotics. Agronomy . Available from: https://doi.org/10.3390/agronomy9070403.

Tsolakis, N., Bechtsis, D., Vasileiadis, G., Menexes, I., Bochtis, D.D., 2021. Sustainability in the digital farming era: a cyber-physical analysis approach for drone applications in agriculture 4.0. In: Bochtis, D.D., Pearson, S., Lampridi, M., Marinoudi, V., Pardalos, P.M. (Eds.), Information and Communication Technologies for Agriculture—Theme IV: Actions. Springer International Publishing, Cham, pp. 29−53. ISBN 978-3-030-84156-0.

van der Merwe, D., Burchfield, D.R., Witt, T.D., Price, K.P., Sharda, A., 2020. In: Sparks, D.L.B.T.-A. (Ed.), Drones in Agriculture, Vol. 162. Academic Press, pp. 1–30. ISBN 0065-2113.

van der Burg, S., Kloppenburg, S., Kok, E.J., van der Voort, M., 2021. Digital twins in agri-food : societal and ethical themes and questions for further research. NJAS Impact Agric. Life Sci. 93, 98–125. Available from: https://doi.org/10.1080/27685241.2021.1989269.

van der Valk, H., Strobel, G., Winkelmann, S., Hunker, J., Tomczyk, M., 2022. Supply chains in the era of digital twins – a review. Procedia Comput. Sci. 204, 156–163. Available from: https://doi.org/10.1016/j.procs.2022.080.019.

Verdouw, C., Kruize, J.W., 2017 Digital twins in farm management: illustrations from the FIWARE accelerators SmartAgriFood and Fractals. In: Proceedings of the 7th Asian-Australasian Conference on Precision Agriculture Digital; Hamilton, New Zealand, 16–18.

Vrchota, J., Pech, M., Švepešová, I., 2022. Precision agriculture technologies for crop and livestock production in the Czech Republic. Agriculture 12. Available from: https://doi.org/10.3390/agriculture12081080.

Zhang, C., Kovacs, J.M., 2012. The application of small unmanned aerial systems for precision agriculture: a review. Precis. Agric. 13, 693–712. Available from: https://doi.org/10.1007/s11119-012-9274-5.

CHAPTER 9

Challenges and opportunities for cost-effective use of unmanned aerial system in agriculture

Chris Cavalaris

Department of Agriculture Crop Production and Rural Development, University of Thessaly, Volos, Greece

Introduction

Unmanned aerial systems (UASs), or simpler drones, evolve to a core element of agriculture 4.0 supporting key aspects for smart agriculture applications, such as crop trait monitoring and precise product applications (Aslan et al., 2022). Their worldwide use is growing exponentially, revealing a broadscale knowledge of their benefits and perspectives for the whole agricultural sector. The provision of objective, timely and high throughput information may assist the effective decision making for critical field operations, including fertilization, irrigation, crop protection, and many more (Lezoche et al., 2020). Furthermore, their introduction into field tasks by substituting conventional machines reveal their integration in the entire farm life cycle (Kim et al., 2019). Their contribution to the adaptation of crop production to climate change and to the reduction of the environmental impact of the agricultural practices is expected to be essential (Pádua et al., 2019; Chivasa et al., 2020).

A characteristic paradigm for the wide acceptance of drones in agriculture is the case of the agricultural sector of China. At the end of 2019, China industry developed more than 170 different types of plant protection UASs and more than 55,000 UASs were operating in an area that exceeded 56.67 million ha. Five thousand (5000) UASs were performing field sprayings only at the region of Xinjiang, the largest cotton-production region in the country comprising of 1.33 million ha (Wang et al., 2022). In Europe, over 150,000 drones are expected to operate in farms by 2035 performing field scouting and material applications (Joint Undertaking, 2016).

Unmanned Aerial Systems in Agriculture
DOI: https://doi.org/10.1016/B978-0-323-91940-1.00009-8

The rate of adoption and spreading of drones in agriculture will largely depend from complementing some organizational, social, and technical challenges, like cost of use and ownership, reliability and confidentiality, heterogeneity of the demands, interoperability, data handling and analyzing dynamics, technical skills disposal, and societal acceptance (Lezoche et al., 2020).

The cost of using drones and the expected economic benefits will be one of the most crucial aspects and will depend on drones' capabilities and technical evolution. The first multirotor UAS was built more than a hundred years ago, but its high cost and poor inherent stability and control hinted its utility until the recent advances in electronic flight control systems that have enabled the inexpensive use of small drones in a variety of civil and agricultural applications (Zhu et al., 2020). Nevertheless agricultural UASs face still numerous technical limitations, such as payload capacity, endurance, communication stability, and security (Kim et al., 2019; Elmeseiry et al., 2021).

The performance of a particular task executed by a UAS depends on the flight speed and ground effective width of the operation. The ground effective width refers to the perpendicular to the direction of travel regarding the performed task. It is usually equal to the distance between two adjacent field passes when a task is performed in parallel lines. There are, however, many limitations compromising the speed and the effective width of the operations. For instance, while performing field spraying the flight speed is limited by the amount of the necessary spray dose as also the desired coverage. Increased speeds also increase the risks of spray drift (Ahmad et al., 2020). Limitations in speed may occur even when performing field scouting operations for remote sensing data collection. The timelapse for image capture and storing of the sensor as well as the requirements for forward overlapping to build the structure from motion models limits the distance intervals between two consequent image captures possessing corresponding limitations on flying speeds (Hardin and Jensen, 2011). The effective width during remote sensing may be high, especially when scouting from a high altitude, but limitations in side-overlapping and reduction in image resolution with the increase of height may also induce restrictions in the operating performance of the drone. User controllable speeds for drones finally are essential for safety reasons as the drone pilot may need to have a sufficient time to react in an emergency situation (Hardin and Jensen, 2011).

Besides the direct functional costs associated with field operations, there are many other factors addressing the cost of utilizing a drone facility in agriculture. Remote sensing for instance implies the collection of a high volume of data requiring increased storage and computational power and some extended time for postprocessing and interpretating the results. High-skilled expert personnel may have to be involved both in data handling and analysis which also imply some considerable costs.

A positive turnover in drone entrepreneurship is that the manufacturing cost becomes constantly lower as more and more companies provide small configurations with low-cost sensors for agricultural and other applications that are accessible to a wider audience (Tsiamis et al., 2019). Moreover, as the drone technology becomes public available and the drone community turn familiar with open hardware designs and software, many low cost do-it-yourself drones are used for low-cost data collection paving the way for wider acceptable commercial solutions (Sørensen et al., 2017; Montes de Oca and Flores, 2021).

Cost components for unmanned aerial systems in agriculture

Using a UAS, regardless of the performed task, always implies some variable and fixed costs (Table 9.1). The variable cost concerns the cost of the operator and its assistants, the cost of the power sources and the cost of energy for recharging, and the costs of repairs and maintenance. It is directly related to its usage and usually estimated at an hourly basis. Fixed cost includes the ownership cost for the drone and its peripherals, the cost

Table 9.1 Cost sources when using a drone.

Variable cost	Fixed cost
Direct costs	
• Personnel cost (drone pilot, assistant) • Power sources cost (batteries, fuel cells, etc.) and cost of energy for recharging • Cost for repairs and maintenance	• Ownership cost (for the drone and its peripherals, like RTK base station, charger, embedded sensors, etc.) • Insurance and fees • Housing and storage • Software licenses
Indirect costs	
• Transportation cost (transfer vehicle, fuels, etc.) • Cost for georeferencing (labor and equipment)	

of insurance and the fees imposed by governmental and other organizations for operating legally, the drone and the costs of software licenses that may be necessary to operate the drone or manage its data. Fixed cost is independent of annual use; however, it is estimated on an annual basis. By accounting the actual annual use, fixed cost can also be expressed at an hourly basis. For example, if the ownership of a drone is estimated at 1000 € per year and its actual flight hours are 20 h per year, then its hourly ownership cost is 50 €/h.

Variable and fixed costs are associated directly to the use of the drone, therefore, they are characterized as **directs costs**. Since drones are always used for a purpose of performing a field survey or an agricultural task, for example, applying a plant care product, there might be some extra costs for complementing the task, like image processing or spray material transportation and preparation in the field. These costs are not directly linked with the drone and that is why they are characterized as **indirect costs**. They might comprise, however, a significant portion of the total cost and they should be also taken into account in the economic studies.

Personnel costs

Becoming a certified drone pilot with particular skills implies some official training and passing through a certification scheme that is specialized according to the category of skills. There are different pilot licenses for instance for different drone sizes, according to their maximum take-off mass (MTOM) or pilot licenses based to the line of sight between the pilot and the drone (visual line of site—VLOS, beyond visual line of sight—BVLOS), special licenses allowing night flights, licenses for dropping materials, and others (Reger et al., 2018). Each category specifies the capabilities of the certified pilots and their limitations. Globally, there is a variety of different classifications and prerequisites among different countries that describe the pilot capabilities and guarantee his capability to fly safety and effectively a drone in the field (Tsiamis et al., 2019). The European Union Aviation Safety Agency (EASA) lately has tried to harmonize the legislation among different member states by introducing a risk-based approach in order to verify certified users with specific capabilities and skills that are able to respond at different operational demands.[1]

A drone pilot therefore may be considered a high-skilled person with also different levels of skills who deserves to get paid accordingly. If, for

[1] https://www.easa.europa.eu/downloads/110913/en.

instance, a drone pilot is requested to perform field spraying, she/he must have a particular license for that operation. Moreover, quite often in complex operations like BVLOS or tasks performed in areas with obstacles, a visual observer might be necessary for assisting the pilot. However, even for single tasks, if the drone has a high mass and needs to be frequently unloaded and uploaded to a truck (a usual case in material applications on small fields), a second person is again necessary to support the operation. Even though this second person does not have to be equally skilled with the pilot, she/he has also to be paid, rising even more the personnel cost. And as personnel cost becomes important, solutions allowing the improvement of the performance of the drone become even more essential. Fuel cell power systems, for example, provide a higher endurance during flight and offset time losses for refiling with energy. Fixed-wings drones are more suitable for long range surveys at a short time provided a suitable runway for take-off and landing.

The personnel involved in a drone operation usually have to spend a lot of time for traveling and perform preparatory or end-up tasks that totally might be greater from the pure missionary time. These tasks must be organized appropriately to minimize their duration and improve the overall performance of a field operation. Time starts "ticking" for a drone application usually one or few days before, by performing some checking for the weather conditions, inspecting the equipment, and requesting the legal permissions for the operation to be performed. On the day of the application, time starts counting from the moment that all the necessary equipment gets uploaded to the transferring vehicle (car or truck) and continues with the time for traveling to the field. When arriving at each new take-off position, all the mission preparations are critical and have to be carried out carefully and methodically in order to ensure successful and efficient flights. Before launching the mission, the drone has to be unloaded from its transferring vehicle, placed in the proper take-off position, unfolded or assembled, supplied with batteries or other power sources, mounted with sensors or filled up with material for plant care applications, programmed by setting up the parameters for its mission, and many more. Field borders might have to be introduced by walking or driving around the field, if the field is visited for the first time, as also some walking inside the field might be necessary to markup physical obstacles, regions of interest, etc. If ground control points (GCPs) have to be used, a person should walk around the field and place them at the proper places and might also have to walk through again at the end to collect them, if they are not going to be left permanently in the field. If a real time kinetic (RTK) base station is going to be used to assist georeferencing, it should be set-up properly in the field, establish communication with the drone and the base station, and allow sufficient time to lock its position from the satellites. Before take-off, a final preflight inspection of all the equipment is always necessary to ensure a successful flight. If more than one sortie is required to accomplish the mission, the drone should return to its launching point, change its power supplies and/or get refilled with material, get a quick inspection, and continue the task with a new sortie. In case the new task or mission has to be performed to some distance away from the previous one, it might be more energy efficient to move to a new launching/landing point. In this case, all the equipment and material should be collected, uploaded to the transferring vehicle, carried, and re-established to its new place. After ending the task, all the equipment has to be disassembled and stored to its cases. If plant care material has been applied, the drone has to be cleaned up thoroughly before packaging. Some time for traveling back is also required and time to unload the equipment to their storage place.

From the above point, it is realized that the personnel cost represents a significant portion of the overall costs (Lampridi et al., 2019a). It should be optimized through the adoption of standard operating procedures and effective routines that minimize the time losses. But the times mentioned above refer mainly to labor associated directly with the field tasks and does not concern the time requirements for tasks performed at a second stage, like postprocessing the images of a field survey, analyzing the results, and extracting data-driven decisions. These times impose a second level of indirect personnel cost that might be even higher than the direct personnel cost, regarding that high-skilled professional staff might be needed to be involved and the amount of time required to accomplish the analysis. According to Reynolds et al. (2019), personnel cost represent a large proportion of the total cost, from 30% to 100% of the cost of vehicles and sensors.

Ownership costs

Ownership cost is a fixed source of cost for drone users meaning that it is present regardless of the annual use. Ownership cost is associated with capital spending and is comprised of two components: (1) equipment depreciation and (2) the cost of interest for the investment. There are various methodologies employed for their estimation (Lampridi et al., 2019b; Sopegno et al., 2016) adapted mainly from classical methods used for cost determination of other equipment, like farm machinery (Hunt, 1995).

The simplest method for estimating the cost of the depreciation is the "fixed method" which allocates equally the deprived value within the whole lifespan. The deprived value equals to the difference between the purchase price and the salvage value at the end of lifespan:

$$D = \frac{P - S}{n}$$

where D = the annual depreciation, P = purchase value, S = the salvage value, $(P\text{-}S)$ = the deprived value, n = the lifespan (years).

Since drones are a rather newbie and fast-growing technology, there is little or no historical information concerning their lifespan and salvage value. Their lifespan, however, is expected to be very short because their rapidly emerging technology results in a quick obsolescence. Assuming a technological evolution similar to that of business computers, a drone would be expected to be replaced every 3–5 years to keep up with technological advances (Jackson et al., 2021).

The salvage value is assumed as the price that a product can be sold at the end of its lifespan. For a drone, it may be assumed equal to 20%–25% of the purchase value presuming that the drone is still functional and may be used on less demanding operations, or sold to amateurs, or parts of it are traded as spare parts.

Besides the ownership cost for the UAS, it is also the cost of its peripherals, such as embedded or mounted sensors, base stations, and chargers, that might be also important. Their technological development follows the advances in drone technology, therefore, they are expected to have a similar lifespan and a similar salvage value.

As mentioned above, ownership costs also include the cost of interest for the investment. This cost exists for all the purchased equipment and should be added to the depreciation costs (D) in order to estimate the overall annual ownership cost. It is usually estimated for the average deprived value:

$$I = \frac{P + S}{2} i_r$$

where I = the annual cost of interest for the investment, i_r = the real interest rate estimated as:

$$i_r = \frac{i_n - g}{1 + g}$$

where i_n = the nominal interest, g = the mean inflation rate.

Cost of power sources

The energy storage unit of a drone is utilized to provide power that may vary a lot during a flight. Power demands are high when ascending or having to perform maneuvers and may be different in a tilted flight, when moving against or toward the wind direction or while hovering. When the drone is used for dropping material (e.g., spraying), the change in payload results in changes in thrust force and consequently changes in energy consumption. Energy is also required for communication with the base station and may be consumed by the onboard sensors and other peripherals that require power for their operation. Due to the extremely variable demands and the complex aerodynamics imposed by different drone types, the estimation of the power and energy demands of the drone is rather complex (Zhu et al., 2020; Zhao et al., 2020; Prior and Bell, 2011). Several models have been suggested recently to provide insights into power demands and energy consumption for small drones. Some of these are quite simple requiring only a few parameters, while others are very complex comprised of multiple interdependent components that provide detailed representations of the forces of flight and drone design. A comprehensive review of such models are presented in Zhang et al. (2021). The models consider the different conditions of movement of the drone (lifting, landing, hover, and tilted flight). They may provide, however, widely divergent results for the same drone delivery operation. Energy consumption rate values for instance may vary by a factor of 3−5. Some models appear very sensitive to the choice of their parameters, or they may respond differently to alternative environmental conditions that cannot be reconciled by using a common set of environmental, drone design, and operating parameters. Advanced component models may provide the most detail information, but are the most difficult to develop and calibrate, given the number of parameters involved. Published results for energy consumption from drone field tests often do not agree with results from the theoretical energy models.

A model based on the blade element momentum theory (BEMT) was used in Zhao et al. (2020) in order to estimate the power consumption of a multirotor UAS powered from batteries for a prescribed flight mission. The authors analyzed the aerodynamic performance of the UAS for lifting, landing, hover, and tilted flight considering the propeller thrust, toque the propeller-induced velocity, the airflow velocity, and the forward speed. Airfoil lift and drag forces were also estimated. Consequently, they accounted for the power transmission losses oriented from iron loss and copper loss on the rotor brushless motors and finally converted the propulsion system power consumption into propulsion system energy consumption (E_{pro}) to calculate the difference between E_{pro} and Li-Po battery energy storage. Some flight test results were performed to validate the above analysis (Table 9.2).

The error between the theoretical and experimental results for power consumption from the above analysis was at the range of 1.1%—27.1% with the greater deviations appearing at lower throttle. The power consumption concerns only the UAS motor power demands and does not account for the energy required for its other internal systems (e.g., inertial measurement unit—IMU, wireless communication, and Global Navigation Satellite System—GNSS). These systems may also absorb a considerable amount of energy and should also be considered in the power consumption concept.

Validation of the model results may be performed by estimations based on actual field data obtained from the missions and on power characteristics from the batteries and the recharging source. These approaches provide a direct insight into actual energy demands accounting also for the consumption of all the peripherals. A simple procedure is described in the below example.

Considering a drone powered by a 51.4 Volt battery with a capacity of 40,000 mAh from which 20% is retained as buffer power for preserving the battery life and for emergency reasons, the drone has 32,000 mAh to accomplish its mission that are equivalent to 32,000/1000 Ah × 54.4 V = 1657.6 Wh or approximately 1.66 kWh. If the endurance provided is 20 min (0.33 h), the average power demand during a flight is 1657.6 Wh/0.33 h = 4972.8 W. The energy consumption can be expressed per ha by dividing the area covered during the 20 min flight (e.g., 1.66 kWh/12.5 ha = 0.13 kWh/ha).

According to the energy resource that is used for recharging the battery, the cost of energy can be further estimated. If for instance, recharging is done back to the office, using the civil electric network, the local costs of electricity should be accounted to estimate the cost. If recharging is made in the field using a fossil fuel-powered electric generator, the specific fuel consumption—SFC (L/kWh) that is the average amount of fuel consumed by the generator per kWh of electricity it generates, should be taken into account. The SFC depends on the fuel type (diesel, gasoline, propane) the generator uses and the load of its engine. SFC decreases while the engine load is increased implying that it might be

Table 9.2 Unmanned aerial system (UAS) power consumption from flight tests.

Movement	UAS total weight 30 kg				UAS total weight 35 kg			
	Throttle (%)	Rotation speed (RPM)	Power consumption (W)	Number of motors	Throttle (%)	Rotation speed (RPM)	Power consumption (W)	Number of motors
Lifting up	50	2250	586.4	8	54	2392	673.9	8
Hover	47	2150	501.0	8	50	2250	586.4	8
Tilted flight—high	54	2392	673.9	4	57	2525	771.9	4
Tilted flight—low	40	1950	343.4	4	43	1980	419.6	4
Landing	37	1790	324.4	8	40	1824	343.4	8

Source: Adapted from Zhao, X., Zhao, J., Wang, X., Ouyang, C., Li, R., Xiong, Y., et al., 2020. A procedure for power consumption estimation of multi-rotor unmanned aerial vehicle. J. Phys. Conf. Ser. 1509, 012015. https://doi.org/10.1088/1742-6596/1509/1/012015.

more efficient to recharge more batteries simultaneously to benefit from an increased load (Mobarra et al., 2022). Typical SFC values for a diesel generator operating at an over 80% load range from 0.260 to 0.280 L/kWh (Mueller-Stoffels et al., 2017). Multiplying the SFC with the amount of necessary energy to recharge the battery (e.g., 0.270 L/kWh $\times$ 1.66 kWh = 0.45 L), the fuel consumption can be estimated. Then the cost of energy can be estimated by accounting for the price of fuel.

Besides the cost of electricity, batteries are consumable devices with a limited lifespan. After the end of their life, they are withdrawn for recycling. A second cost of use orients therefore from the gradual loss of their storage capacity. Batteries are characterized by their lifecycles, that are the potential number of charging and recharging cycles and age as the number of cycles increase because they lose their capacity. The number of cycles depends largely on the proper battery use and maintenance. Charging and discharging outside their nominal voltage window causes physical degradation and performance loss. Typical lifecycles for Li-Po and Li-ion batteries range from 500 to 2000 h (Dündar et al., 2020; Liang et al., 2019). A prolonged life cycle for a battery pack implies a lower cost of use (Kim and Lee, 2020).

For drones powered with fuel cells, the consumed fuel is hydrogen, as the oxygen is pumped from the atmosphere. Fuel cells are of the most efficient devices for extracting power from fuels (https://www.energy.gov/sites/prod/files/2015/11/f27/fcto_fuel_cells_fact_sheet.pdf). The fuel cell efficiency is defined as the electricity produced per kg of hydrogen consumed. For a hydrogen-fed fuel cell, it is directly proportional to cell potential (Barbir, 2005):

$$\eta = V/1.482$$

where η = the efficiency and V = the cell potential in volts.

Tests performed in Wang et al. (2021) to evaluate a small drone performance using different types of fuel cells revealed that in order to supply the drone with an average power of 1.46 kW, an average hydrogen consumption of 8.5 g/h was required. The cost for using hydrogen as a fuel therefore can be estimated by taking into account the commercial prices for hydrogen. However, since hydrogen is provided in compressed tanks, that for UASs are partly made of carbon fiber, the cost of the hydrogen storage tanks, if they are not reusable, may be significant and should be also considered. According to Belmonte et al. (2017), the main drawback of fuel cell-based systems is, at present, represented by their high cost, which is about double that of a battery-based system of the same size and for the same application. Reducing cost and improving durability are the two most significant challenges to fuel cell commercialization.

In hybrid systems, involving fuel cells and batteries, the cost of using a battery should be also considered in the way mentioned above.

For drones with internal combustion engines, an average hourly gasoline fuel consumption can be accounted for 1.246 L/kg of MTOM. This is the average value estimated from hourly fuel burn data for small private aircrafts (Shen et al., 2021). Considering the price of fuel, the cost of the power source can be estimated consequently.

Costs for repairs and maintenance

The use of drones imposes some risks emerging from potential accidental operations, subsystems failure or damage caused by unpredicted factors like unexpected wind gusts, collision with another aircraft, and others. Commercial drones contain various safety systems, like ultrasonic sensors and light detection and ranging (LiDARs), that help prevent accidental incidents improving safety even for less-skilled operators as also systems informing the user about the operational ability of the drone. But beyond all the preventive measures that may be taken into place for ensuring a safe mission, the normal use of a drone always imposes a physical degradation of its components that finally may end up with a light or more severe deterioration of the drone's functions. The light deterioration of some functions that do not result on operational inability and mission abortion are called a "soft failure." Moderate degradations may also occur that lead to mission abortion without, however, causing severe damages. These types of failures result in unscheduled downtime due to repeated missions that in turn, impose a serious increase in the operational cost. More severe failures may result in heavy and expensive damage with a rather small probability of repairing. In that case, considerable time may be lost until the drone is repaired or replaced. In most serious cases, catastrophic damage may result in drone abortion due to inability for a repair for technical or financial reasons. Such a case may lead to a serious loss of the expected lifespan resulting in considerable higher ownership costs. Litigation incidents related to third party damages or injury may have another negative impact.

The probability that a system, subsystem, or part perform its specific function in a preestablished time, under preestablished conditions is called

Table 9.3 Reliability and failure rate parameters for a commercial type of drone.

Subsystem	FIT (F/10^6 h)	MTBF (h)
Ground control system	2.00	500,000
Mainframe	2.77	360,984
Power plant	9.94	100,603
Navigation system	9.41	106,269
Electronic system	5.01	199,600
Payload	1.10	909,090
FIT (total)	30.23	—
MTBF (R_{total})	—	33,080

Source: Adapted from Petritoli, E., Leccese, F., Ciani, L., Roma, S., Navale, V., 2018. Reliability and maintenance analysis of unmanned aerial vehicles†. Sensors 18, 1—16. https://doi.org/10.3390/s18093171.

reliability (Petritoli et al., 2018). The opposite of reliability is failure rate, which describes the number of failures that are possible to occur over an established time. Reliability may be described by the mean time between failures (MTBF) (usually expressed in hours) while failure rate may be described by the failure in time (FIT), that is, the number of expected failures in one billion device-hours of operation. Table 9.3 presents reliability and failure rate parameters for the common subsystems of a commercial drone. Many drone manufacturers deploy a duplication of the most critical or vital subsystems in order to enhance reliability and prevent critical failures, but this is not always an ideal solution as it increases the cost, the weight, and the complexity of the system. Despite the significant technological progress in drone reliability, in recent research it was highlighted that drone users complain mostly for functional errors (27.9%), device compatibility (16.8%), cost (16.2%), and connection/synchronization issues (15.6%) (Klauser and Pauschinger, 2021).

Periodical maintenance of the drones is essential for preventing system degradation and restoring functionality. Improper maintenance or insufficient repairs could result in serious drone damage, but even soft failures may impose a poor performance resulting in an increased cost of use until the correcting maintenance action is performed. Quite often, soft failures are the underlying cause for heavier damage. A repair action is required thereafter to restore the problem which may impose a higher cost. Therefore preventive maintenance actions are essential even if they introduce some extra cost. This cost may be optimized if the optimal preventive maintenance intervals, that do not exceed the repairs threshold, are determined. For that scope, the failure rate of all the critical systems and subsystems of the drone should be accounted (Petritoli et al., 2018). Generally, approximately 5%—10% of the purchase price of the drone can be associated with typical annual costs for repairs and maintenance (Jackson et al., 2021).

Insurance and fees

Drone insurance is liability coverage intended to protect the owner or the operator against claims for property damage or bodily and personal injuries caused by the drone. Property damage covers a drone accident resulting in a damaged property of others. Bodily injury apply to the medical expenses of people injured by a drone crash while personal injury coverage applies to libel, slander, and invasion of privacy or other psychological-based claims (Jackson et al., 2021). The liability coverage against damages and injuries to third parties is a legal obligation for all professional drone users. The provided coverage is compliant with minimum national and international regulations, like the EC EC785/2004 regulation of the European Parliament and of the Council. Optionally, the drone owner may request hull coverage that covers physical damage of the drone and may also include damage caused to valuable payloads such as a hyperspectral camera, LiDAR, or another types of sensors. Theft, malicious damage, and system hacking are also some additional coverages that may be provided to drone users by the insurance companies.

Insurance for agricultural drones is essential because it protects professionals from financial ruin as accidents are likely to occur even in an open field. The cost for drone insurance can vary largely depending on the type of drone, its weight, the environment, the performed tasks, the amount of liability coverage, and the additional coverage requested by the owner. Since the professional use of small drones is rather new, insurance companies today are challenged to address unknown risks, without having a database of claims to use as a reference. A policy based on drone weight and activity rating appears to be a simple approach. Annual policies for liability coverage usually start at around 500 € but may vary significantly among different countries. Drone users, however, may tailor their coverage policy according to their needs and request an annual plan, purchase coverage by the month, or even ask for a daily plan.

In some countries, annual fees are also required for professional users to retain their license. This is an additional expense that should also be encountered while estimating cost of using a drone.

Software licenses

Most flight planning applications for drones are free. Automated image capture and video registration is also a free available utility for most of the apps. Image postprocessing software, however, and functions related to image classification, and field zonation for precision agriculture applications require an annual or monthly subscription. Some companies also provide the opportunity of a one-time fee for a perpetual license. Holistic schemes encompassing image processing, field diagnosis, creation of prescription maps and corresponding flight planning are also available at a monthly, annual, or perpetual policy from some drone manufacturers and software developers. On the other hand, open source software packages and algorithms like OpenDroneMap, QGIS, SAGA, and OpenCV classification algorithms provide the opportunity to develop replicable workflows for low cost applications (Mattivi et al., 2021).

Housing and storage

Small drones generally occupy a limited space, especially when they are folded so their storage does not require a large storeroom. Some extra space, however, is required for its accessories like sensors, charger, power generator, and RTK GPS. If recharging of the batteries is performed on the storeroom, some special place, with particular safety specifications is also a prerequisite. The room should also have established an automatic fire extinguishing and suppression system to avoid a potential fire from accidental explosion of a battery. Heating and cooling might also be necessary to retain the room temperature between specified limits in order to ensure prolonged battery life. Since the stored equipment including the drones, their peripherals, and sensors are of high value, a security antitheft system might also be necessary to protect the property. Housing and storage, therefore, of a professional drone entail specific conditions that are far beyond a simple warehouse, increasing the cost of ownership. In the case of a bigger company, operating a fleet of drones, the storage space, and the safety demands are much higher increasing further the cost. In that case, however, the space may be organized more efficiently, reducing the storage cost per single drone. The cost is estimated on an annual basis comprising from the building construction costs and the costs of interest or opportunity, amortized over the expected life of the building as also the real estate taxes, insurance, and repairs and maintenance expenses to retain the building functional and the cost for having a security service. These costs are shared by the number of drones stored in the room.

Transportation costs

Drones should be transferred to the operational area for performing their tasks. Mini drones equipped with small sensors for field surveys can be carried rather easily even with private cars, but when the needs imply a higher payload, a bigger drone with increased dimensions may be required. The accessories may also require considerable space. In the case of field spraying, for instance, a professional operator should carry to the field a drone, that even if it is folded may occupy over half a cubic meter as also a set of batteries, a power generator, a real-time kinetic global positioning system (RTK GPS), a tank of water that may be over 200 L, packages of plant protection products, accessories for preparing the mixtures, and other. In such a case, a professional vehicle is required for the transportation that has some fixed costs associated to ownership and interest or opportunity that are amortized over its lifetime, as also cost for insurance and fees, variable cost associated to repairs and maintenance, cost of fuels, and personnel costs for the driver. In the case of professional drone services, the vehicle may be employed solely for drone purposes while in the case of a farmer, it may be a multipurpose vehicle covering variable needs in the farm. In the first case, the whole cost is attributed to the drone services while in the second, it is addressed according to the time allocation in drone transportation in relation to other uses.

Despite their direct or indirect origin, the variable and fixed costs are expressed in an hourly basis but for the validation of the profitability of the performed task they should be expressed at an acreage basis. The conversion from an hourly to an acreage basis implies the estimation of the actual performance of the task (for instance ha/h) which is highly related to the capabilities of the drone, the endurance of the power system, and the restrictions imposed by the specific tasks. Different types of drones (e.g., fixed wings, multirotors) present different capabilities making them suitable for alternative tasks. But one of the greatest challenges today in drone performance is their power supply. Except for the case of tethering, drones have to bear with their power sources (batteries, fuel cells, etc.) but due to restrictions in weight and size, their bearing capacity is limited resulting in limited autonomy. The limited autonomy imposes significant time losses for "re-fueling" with energy the power supply system resulting in poor performance and increased direct costs. Similarly, restrictions imposed in indirect tasks like tagging a survey area for ensuring correct georeference or postprocessing the material of a hyperspectral survey may

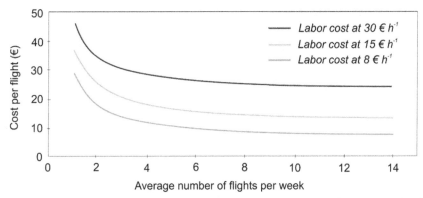

Figure 9.1 Change of total annual cost per flight with annual use for three levels of personnel cost. *Adapted from Jackson, J., Ladino, K., Abdulai, G., Engineering, A., 2021. Decision Aid for Estimating the Cost of Using a Drone in Production Agriculture, Kentucky.*

also result in poor performance and a high functional cost. All these restrictions should be considered carefully in order to identify the most cost-effective solutions for obtaining the expected results.

Fig. 9.1 presents the change in the total annual cost per flight as the number of flights performed per weak increases (Dolias et al., 2022). There is an inverse relationship between the number of flights per year and costs per flight. The cost per flight decreases with an increasing number of flights because the fixed costs of ownership, insurance, and fees are allocated over more flights. As the number of flights conducted per year increases, the total annual cost increases, and this increase in total cost is controlled primarily by the personnel cost.

The College of Agriculture Food and Environment of University of Kentucky developed a simple computational decision aid tool for estimating the cost of use and assisting professionals and farmers prior purchasing UAS for monitoring agricultural ecosystems (Jackson et al., 2021). The tool accounts for the equipment ownership cost, cost of certificates and licenses, personnel cost, cost for utilities, cost of purchasing or using associated software, insurance cost, and cost of maintenance and repairs of the system. Different cost scenarios can be examined in terms of the onboard camera specifications and image capture settings like required spatial resolution and flight altitude, drone speed and necessary image overlap, parameters associated directly to the area covered during a single flight, and the time for postprocessing analysis.

Cost-effective unmanned aerial systems applications

Drones are used for two main scopes in agriculture: surveillance, and material applications. Each case has its own limitations and challenges for achieving a sustainable and cost-effective use. Surveillance for instance besides effective flight planning, imposes the need for geometric accuracy, data handling, and image processing, that are directly related with the cost of these tasks. Material applications, such as pesticides, fertilizers, and seeds, require accurate placement, avoidance of drift transfer of large quantities, and other operational issues.

Surveillance

Photogrammetric products acquired by drones today are utilized broadly for classifying vegetation types, assessing plant vigor, estimating leaf area, predicting photosynthesis, identifying biotic and abiotic plant stresses, and many more. Until recently, many of these tasks were based on ground observations or were performed with ground portable devices that implied a high human effort and an increased operational cost. Even cheap or free available mobile phone apps, developed recently for assessing various plant traits, present a very low scalability that does not allow a wide scale interpretation of their readings in order to address effectively the spatial heterogeneity. But the potential of using aerial products from UASs opened a new era for monitoring open field agriculture with a high frequency and resolution at an affordable price that permits interventions at a farmer level for improving productivity and net profit (Losè et al., 2020). Moreover, the recent development of small and cheap UASs carrying on-board low-cost imaging sensors, along with the development of effective image processing software allowed the wide integration of UAS-based photogrammetry into precision agriculture applications.

The most competitive technology to UAS remote sensing today is satellite-based imagery provided for free, including the LandSat and the ESA Sentinel products. These products can map large areas but quite often, due to their low spatial resolution, fail to address specific needs in agriculture, like identifying the canopy of a tree or a vine row in order to access its geometry or recognizing spatial variability in small fields. Moreover, satellite observations are often restricted by cloud cover resulting in missing important information. Therefore the profitability of using UASs instead of satellite data depends largely on the cost of use and the importance of getting timely information with a higher resolution.

According to Yi et al. (2021), by using a drone-based multispectral sensor to map the normalized difference vegetation index (NDVI) and implement a variable rate fertilization scheme in grains, the UAS solution do not appear profitable because the average cost per hectare is higher than the average economic profit resulting from its use. In such a case, a lower resolution satellite acquired imagery may be a more suitable solution. On the other hand, drone-acquired images can be highly profitable when they are used for higher value crops or high added value applications like precision spraying or foliar fertilizer applications in vineyards (Yi et al., 2021). The size of the surveyed area is also important from an economic point of view. A recent study in Switzerland where farms are rather small, suggests that the cost of using a drone for scheduling variable rate fertilization should not exceed 6.5 € per hectare and per year if it is to be more profitable than a low-resolution technology like field sampling with an N sensor, or not more than 4.5 € per hectare per year compared to the use of a 10 × 10 m satellite image (Späti et al., 2021). The study also suggests that the net profits increase when a high-resolution technology is applied to fields which exhibit higher spatial heterogeneity of soil conditions and lower spatial autocorrelation. In Matese et al. (2015), it was identified a break-even point of 5 ha, below which, the adoption of UAS imagery is more advantageous from satellite or aircraft imagery.

The use of low cost RGB image sensors in small UASs is another opportunity. These kind of sensors are able to access some important plant traits like green matter coverage, plant geometry from structure of motion processing, and other (Schirrmann et al., 2016). High resolution RGB images combined with machine learning techniques may provide insights to even more advanced plant traits like leaf chlorophyll (Guo et al., 2020). RGB-derived vegetation indices like visible atmospherically resistant index (VARI) (Feng et al., 2020), triangular greenness index (TGI) (Feng et al., 2020), and many more are often utilized for estimating crop nutrition and other traits that sometimes perform as well or even better than indexes derived from more advanced and expensive sensors like multispectral and thermal cameras (Gracia-Romero et al., 2019). According to a study examining the reliability of consumer grade RGB cameras mounted on drones, there are three important preconditions that have to be fulfilled in order to get correct results: (1) the ambient light conditions should be stable during image acquisition, (2) the angular variation in reflectance is taken into consideration, and (3) testing of the orthorectification software to evaluate how it handles bidirectional reflectance

(Rasmussen et al., 2016). In Schirrmann et al. (2016), it is stated that low-cost RGB imagery from UASs can strongly support farmers in observing biophysical characteristics, but it fails to address the nitrogen status within wheat crops.

According to Matese et al. (2015), there are three special sources of costs during the pipeline of surveillance with drones:
1. Acquisition costs,
2. Georeferencing and orthorectification costs, and
3. Costs associated with image processing.

A surveillance procedure therefore should be organized properly to eliminate each individual cost in order to benefit from the added value of the derived information.

Acquisition cost

Acquisition costs refer to all the expenses related to getting the raw images from the field including all the costs for organizing and conducting the acquisition campaign. They include all the costs described above comprising the ownership cost, cost of personnel, cost of energy, transportation cost, etc. The cost concerns the drone, the onboard sensors, and the peripherals.

A key parameter to reduce the above cost categories is reducing the time spent in the field. Optimizing the flight time by selecting efficient flight paths, take-off locations, and recharging plan is crucial for eliminating the acquisition cost. Efficient flight planning aims to determine an optimal path that allows the completion of a mission in a minimum time and provides an effective use of the limited capacity of the drone power supply system. Several efforts have been made in the development of energy aware path planning algorithms to extend the size of the area a drone can cover (Tamayo et al., 2020; Benos et al., 2021; Dolias et al., 2022). Heuristic, multiobjective, and other type of algorithms have been proposed for selecting the path with the balanced score from few turns, few revisits, completed coverage, and minimum traveled distance (Campo et al., 2020). A drone requires more energy when making turns than going straight (Tamayo et al., 2020). For large surveys, in-flight recharging, swapping, and wireless recharging techniques may have to be implemented (Subramaniam et al., 2021). Since ground stations are also expensive, effective placement in the field, according to the endurance of the drone, is important for minimizing the cost of the operation. Dividing the area into discrete squares or cells are some approaches proposed by researchers to minimize the total flying distance and achieving low-cost drone coverage of large areas (Yi et al., 2021).

Georeferencing and orthorectification cost

Obtaining georeferenced and orthorectified imagery during field surveys is essential and crucial especially when multitemporal studies are employed or when the images are going to serve as baseline for deriving accurate application maps, such as spraying specific targets and applying variable rate fertilization. Global navigation satellite systems (GNSS) data by itself, without correction, may provide a maximum accuracy of about 1 m but may deviate to a few meters in cases of poor GNSS receiver connections.

Georeferencing and orthorectification costs include the costs for purchasing or constructing artificial geotagging material, for example, GCPs, the cost of complementary devices, for example, GNSS or a RTK base station and the man-hours needed to spend for setting up all the auxiliary equipment.

Despite the georeferencing process having achieved a high level of automatization nowadays, there is still considerable effort required to obtain an accurate imagery product. At least four GCPs are necessary in order to effectively georeference a UAS-obtained orthomosaic product but the use of a higher number of GCPs is generally advised to introduce redundancy and to better estimate the camera interior orientation parameters (Losè et al., 2020). A high number of GCPs is also necessary for building orthomosaics from similar images captured over uniform fields with limited spectral variation (like the case of a big cereal field with no discrete areas of variability in vegetation).

The GCPs have to be placed manually over the survey area and their position should be geo-tagged using traditional topographic techniques or by means of GNSS receivers. This task may be one of the most time-consuming during the field scouting process and may be extremely difficult in the case of surveying hardly accessible areas. The GCPs may need to stay in place for a longer period in case a time series set observation has to be employed and may need to be cleared or maintained regularly during a survey period to remain discrete on the aerial images. In some cases, natural and permanent points of interest, visible on the aerial images may also be utilized to eliminate the needs of artificial targets but their positions should be marked at least once at the beginning.

Besides field tagging prior the image acquisition stage, the use of GCPs in the data processing phase may be also time consuming since an operator needs to manually identify the points on the images in order to link the map coordinates with the image ones (Losè et al., 2020). Some solutions to speed up the postprocessing procedure propose the use of automatic or semiautomatic recognition based on computer vision techniques (Losè et al., 2020). There are also some strategies proposed to optimize the field preparatory task, like using coregistration approaches among multitemporal datasets and optimization of the number of GCPs (Losè et al., 2020). More recently, the availability of high accuracy, survey-grade GNSS receivers in UASs, utilizing double frequencies and multiconstellation techniques, along with information concerning the drone heading, roll, pitch, and yaw retrieved by the onboard IMUs, has made possible the direct georeferencing of the images without the need of using any GCPs (Cucci et al., 2017). Direct georeferencing solutions may reach, in theory yet, a positional accuracy of even few centimeters (Losè et al., 2020). There are two main technical approaches followed, the RTK technique and the postprocessed kinematic (PPK) (Losè et al., 2020). In the RTK approach, the camera positions measured by the onboard GNSS RTK receiver are corrected in real-time by utilizing additional georeference information sent by a stationary base GNSS receiver placed in the field, or by a virtual GNSS station created by a network of continuously operating reference stations (CORSs), a solution known as NRTK. In the PPK approach, the correction of UAS positions recorded by the onboard GNSS receiver during the flight and the estimation of the camera positions are carried out after the data acquisition phase, with suitable software. The RTK approach is faster providing real-time, accurate positioning of the drone but requires constant and uninterrupted communication between the drone and the base station restricting its applicability in visual line of sight VLOS missions. In case of a momentum loss of communication and until its re-establishment, it may likely end up with gaps in the dataset resulting in a significant reduction of the quality. The PPK approach imposes a way simpler setup, it is more accurate, but needs more computational power and impose extra time in the postprocessing procedure (Padró et al., 2019). Nevertheless, it is much preferable than having to repeat the mission in case of insufficient data quality from an RTK survey.

Image processing costs

Image processing cost include the cost of use of the photogrammetric software, the cost of the hardware to run the software, the cost for storage of primary images, metadata and the processed data, and the cost of the human effort involved in the data process. The image datasets from a field survey are difficult to be analyzed because they are voluminous, complex, heterogeneous, often plagued with errors, and only can be handled with up-to-date scientific and mathematical tools (Reynolds et al., 2019). The recent development of commercial packages including data acquisition tools and specific image processing pipelines have relieved the bottleneck of data analysis providing the chance for routine tasks to be accomplished automatically in a high-throughput fashion (Reynolds et al., 2019). Public consortia develop also flexible analytic workflows, which are interpretated by the scientific community through an "open science and open source" approach providing constantly new solutions for effective and fast data processing even for material captured from low-cost devices.

Field surveys regularly produce a huge number of images that must be processed in photogrammetric software to provide a valuable product for end-users. Depending on the flight attitude and the image overlap, a field survey with a four-band multispectral camera may produce over 500 images for a 10 ha field. Postprocessing of such a high amount of data implies the use of high computational power and computers with a higher cost. The minimum RAM requirements for PIX4D Mapper (http://www.pix4d.com) to process medium size projects containing 100−500 images is 16 GB while for large projects with over 1000 images, 64 GB is required. Moreover, the raw data as well as the final photogrammetric products have to be stored for back-up purposes and potential farther analysis. A single project containing 500 raw images at 1.2 MP resolution requires over 2 GB of storage space. A hyperspectral survey over 1 ha may produce over 10 GB of data. The volume of data becomes enormous as thousands of data points and images collected in multiple experiments, together with necessary metadata and this is a great challenge for cost-effective surveillance in agriculture with high resolution drone sensors (Hardin and Jensen, 2011).

Increased area coverage with less or lower resolution images results in reduced postprocessing times but the quality of the final product is poorer and may become insufficient for detecting some plant traits impairing the decision-making capabilities. A cost tradeoff exists between the quality of images and the time for image analysis. The establishment of a standard image acquisition protocol including specifications for imaging conditions in terms of illumination, wind, height, side-overlap, image resolution, lens adjustments, and other parameters may improve image homogeneity and clarity, and conserve human effort for manual image checking and correction and reduce computational times and demands (Reynolds et al., 2019). Overall, the cost of image analysis in survey projects is the most underestimated and may represent 10%–20% of the cost of image capture (Reynolds et al., 2019).

Aerial material applications

Using drones to apply plant care products in the field, including pesticides, fertilizers, seeds, and other agricultural commodities, is another challenge for the UAS technology. The main restriction in such applications is the high payload that has to be carried by the drone, increasing its size and reducing considerably the endurance of its power supply system (batteries, fuel cells, etc.) (Subramaniam et al., 2021). A small drone powered by a Li-ion battery and carrying a light sensor for surveillance purposes may have an autonomy of 25 min and even more while a drone with a 20 L tank performing field spraying and powered also by Li-ion batteries is hardly to achieve an over 15 min autonomy. Payload restrictions result also in some changes in common agricultural practices. Using condensed products or performing targeted applications are more suitable for UASs. The dilution rate of a plant protection product might be 10–30 times lower in the drone tank compared to the rates used in conventional ground field sprayers.

Material applications are performed mainly by multirotor UASs because they can take-off and land vertically, having minimum space requirements. A drone applying material may have to land for refiling the tank and changing its batteries multiple times during a field task. To avoid nonworking routes, landing may have to be done at different places. A fixed wing, carrying a high payload, requires an appropriately reformed runway to take-off and land and this is not always available in the field and especially at multiple points. As drone-based applications, however, are expected to become regular in the future, it is also anticipated that farmers will reform their land accordingly by establishing permanent take-off and landing points and by creating paths for moving from one point to another. These permanent take-off and landing points should be allocated in the field in a way to support an efficient flight planning that aims to maximize the covered area during the limited endurance provided by the power supply system of the drone. Minimizing dead routes and turnings are critical factors to be considered as also the physical restrictions imposed by field geometry and terrain, natural obstacles, and crop orientation. Retaining a visual line of site with the drone all over the missions corresponding to a specific landing and take-off point is also crucial.

Many agriculture drone manufacturers provide online information on the remote controller about battery power level and the remaining payload to support operators on their decision whether to return the drone for a refill and batteries replacement. However, during aerial material applications, the endurance of the power supply system and the periods for refiling should be synchronized in order to minimize the needs for landing and take-off and conserve energy and time. Time losses from frequent take-off and landing for refiling with spray liquid and exchanging batteries, during pesticide applications for instance, may increase the operational cost by more than 40% (Xu et al., 2020). The synchronization of times for replacing batteries and tank refiling may be achieved in various manners like adjusting the flight speed, changing the attitude, regulating the dose, or altering the dilution rate of the plant protection products (PPP). For instance, a DJI T30 drone with a 30 L tank has a hovering endurance of about 13 min with a mean payload, operating on no-wind conditions. Considering a flight height of 1.8 m that provides an effective spray width of 5 m, a flight speed of 2.5 m/s and a field efficiency during flying of 0.9 (10% time losses for turnings, avoiding obstacles and moving to the starting and landing points, but without considering the times being on land), a field performance of 4.05 ha/h is estimated. The area sprayed during a 13 min flight though is about 1 ha. If the drone is performing field spraying of a PPP at a dose of 3 L/ha diluted at a rate of one-fifth on water, it applies a dilution volume of 15 L/ha. Therefore the drone has to be filled with 13.2 L of spray dilution to spray 15 L/ha at 0.88 ha in 13 min. But using a lower payload (13.2 L instead of 30 L) reduces the power requirements and increases the battery endurance. There is an inverse power relation between the maximum hovering time and the gross take-off weight of a multirotor drone as depicted in Fig. 9.2 (Vu et al., 2019). By using such a relation for the DJI T30 drone and performing optimizing computations, a higher hovering endurance of 15.1 min is estimated and the potential sprayed area is calculated at 1.02 ha (Table 9.4, Case 1). The performance of the operation may be improved by altering some spraying parameters within permittable limits. A potential increase of flight speed, for example, from 2.55 to 3.33 m/s could increase field capacity at 1.27 ha per battery by using 19.05 L in the tank (Table 9.4, Case 2). The potential also to use a higher spray altitude to increase effective spray width from 5 to 7 m could

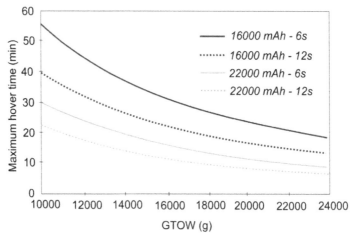

Figure 9.2 Calculated maximum hover time versus gross take-off weight of an agricultural quadcopter for several of batteries configuration. *Adapted from Vu, N.A., Dang, D. K., Le Dinh, T., 2019. Electric propulsion system sizing methodology for an agriculture multicopter. Aerosp. Sci. Technol. 90, 314–326. https://doi.org/10.1016/j.ast.2019.040.044.*

Table 9.4 Alternative scenarios to improve drone efficiency by synchronizing battery change and tank refiling periods of a 30 L spraying drone.

	Case 1	Case 2	Case 3	Case 4
Dose (L/ha)	3.00	3.00	3.00	3.00
Dilution rate (L PPP/L of water)	1/5	1/5	1/5	1/6.5
Speed (m/s)	2.50	3.33	3.33	3.33
Effective spray width (m)	5.00	5.00	7.00	7.00
Field efficiency	0.90	0.90	0.90	0.90
Hovering endurance (min)	15.1	14.1	12.9	11.9
Capacity (ha per battery)	1.02	1.27	1.62	1.50
Tank fill (L)	15.28	19.05	24.36	29.24

increase further the field capacity to 1.62 L/ha by using 24.36 L in the tank (Table 9.4, Case 3). If a higher volume is preferable to improve crop coverage, there is a chance to use an 1/6.5 dilution rate that would decrease slightly field capacity to 1.5 ha per battery but would take advantage of almost the full capacity of the spray tank by adding on it 29.24 L of dilution (Table 9.4, Case 4).

The above computations are proximal since actually the mass of the payload during spraying is gradually decreased over time changing the power requirements for hovering, as depicted in Fig. 9.2. The current load is not connected by a linear function to the operational time, so using a mean payload weight leads to a proximal estimation of the hovering endurance. Exact models should be provided ideally for each type of drone to access hovering and battery endurance and consequently the appropriate payload in the tank.

For material applications at big farms with large fields, swarms or fleets of drones may be used to increase performance, conserve time, and eliminate the personnel cost since one or few operators can handle simultaneously multiple drones that cover a larger area. In swarm UAS systems, a set of UASs work together to achieve a specific goal. Each UAS has a small mission which is a part of a bigger mission (Elmeseiry et al., 2021). Experimental results indicated that the performance of the swarm system is significantly superior to the single drone system (Ju and Son, 2018). Multiple UASs may require more time to set up and impose an extra ownership cost but a reduction in working time of about 18%−50% together with the reduced operator cost justify the investment, especially when used in large areas.

A common challenge for swarm UASs is the appropriate assignment of tasks to each drone and their sequence in order to optimize the overall field performance. Multiobjective and genetic algorithms have been proposed for that scope that minimize the nonoperation flight distance and the operational cost (Xu et al., 2020; Ivić et al., 2019; Hong et al., 2021). The algorithms determine the waypoints and the flight paths for each UASs and take care to prevent overlapping of flight areas and drone collision. They are capable to improve the performance by 35%−65% and also improve accuracy and mitigate overspraying in the range of 3%−8% (Ivić et al., 2019).

Current state and future perspectives

UAS technology is developing and improving rapidly. Agriculture is expected to be the second most important sector after infrastructure to be benefited from the introduction of UASs (Losè et al., 2020). New products appear every year into the market to cover the largely diversified demands of the farmers. In 2015 in China, agriculture UAS spraying were registered at 667,480 ha while 4 years later, in 2019, the annual cumulative spray area was more than 33 million hectare (Xu et al., 2020). In 2015 plant-protection UASs had a payload up to 10 kg while today, models of 30 and 40 kg or even higher are available. Low cost small drones carrying mini sensors along with the development of user friendly photogrammetric products employing simple general user interfaces and routines have made widely available a field survey technology that before some years were utilized only by a small scientific community (Jang et al., 2020). The development of cost-effective tools based on open-source geospatial technologies will make feasible the implementation of field scouting even in small farms. Improvements in batteries power and energy density and the much promising technology of fuel cells is expected to break the barrrier of limited endurance and improve the economic performance of the applications. Innovations in the propulsion system like the double-rotor has made possible an improved drone design that enhances the penetration of PPPs into the canopy of the crops. The opportunity to operate multiple UASs in swarms is opening the gate for the introduction of task-performing UASs into large farms.

Numerous case studies and experimental data reveal the economic feasibility of using UASs for performing field tasks or deploying sensors to provide information for smart and precision agriculture applications. A study in vineyards proved that using site-specific spraying based on canopy size maps obtained by drone sensors could reduce pesticides use by up to 80% (Andújar et al., 2019). In another study, it highlighted the opportunity to perform low-cost and high-efficiency UAS spraying application for olive and citrus orchards. Compared with conventional ground spraying, the UAS application presented a 7 €/ha lower cost (Martinez-Guanter et al., 2020). A study in aphid control in cotton revealed that the operation efficiency of the UAS was increased threefold, and the amount of pesticide was reduced by one-third. Especially for cotton, spraying with ground-based sprayers during the later growth stages introduces significant mechanical damage to the plants. Instead, using UASs allows frequent applications to control pests without fearing mechanical damages and consequent yield losses. Besides spraying, there are many other tasks that have potential to be implemented by drones. Rice sowing for instance has proved to be time conserving and labor saving compared to traditional methods using mechanical transplanting and sowing (Wang et al., 2022). Weed detection and control is another case. Low-cost UASs combined with open-source software seem suitable for weed detection compared to field surveys and other conventional techniques (Mattivi et al., 2021). The information can be used to create prescription maps for performing site-specific herbicide spraying beyond a decision threshold or variable rate applications.

Agriculture UAS applications are expected to accelerate as the benefits will become widely acknowledged. UASs will affect the agricultural sector not only in a practical sense by changing existing farming practices and forcing adaptations to the crops and fields, but also symbolically, by bringing particular values and imaginaries and altering the society's perception on farming (Klauser and Pauschinger, 2021). Farmers who use drones consider themselves early adopters who are seeking practices to reduce production costs and/or increase yields but also recognize that special skills and effort required to fully leverage and implement a profitable system (Bir and Mintert, 2019). On aware of that, many agricultural drone manufacturers adopt a sales and services policy by setting up professional pest-prevention organizations or plant-protection companies to undertake plant-protection services directly by providing big data support and deploy management services, supervise the quality of the operation throughout the whole process, and standardize the operation process and parameters (Wang et al., 2022). A recent inventory revealed the six top benefits of the drone technology, as perceived by the farmers as follows: it saves time, makes pesticide application easier, requires less labor, enhances pesticide application effectiveness, is superior to other pesticide application methods known to the farmers, and reduces the negative impact of pesticides on the environment (Annor-Frempong and Akaba, 2020).

References

Ahmad, F., Qiu, B., Dong, X., Ma, J., Huang, X., Ahmed, S., et al., 2020. Effect of operational parameters of UAV sprayer on spray deposition pattern in target and off-target zones during outer field weed control application. Comput. Electron. Agric. 172.

Andújar, D., Moreno, H., Bengochea-Guevara, J.M., de Castro, A., Ribeiro, A., 2019. Aerial imagery or on-ground detection? An economic analysis for vineyard crops. Comput. Electron. Agric. 157, 351−358. Available from: https://doi.org/10.1016/j.compag.2019.010.007.

Annor-Frempong, F., Akaba, S., 2020. Socio-Economic Impact and Acceptance Study of Drone-Applied Pesticide on Maize in Ghana. Wageningen, The Netherlands.

Aslan, M.F., Durdu, A., Sabanci, K., Ropelewska, E., Gültekin, S.S., 2022. A comprehensive survey of the recent studies with UAV for precision agriculture in open fields and greenhouses. Appl. Sci. 12. Available from: https://doi.org/10.3390/app12031047.

Barbir, F., 2005. Fuel cell electrochemistry. In: Barbir, F. (Ed.), PEM Fuel Cells. Academic Press, Burlington, pp. 33−72. ISBN 978-0-12-078142-3.

Belmonte, N., Luetto, C., Staulo, S., Rizzi, P., Baricco, M., 2017. Case studies of energy storage with fuel cells and batteries for stationary and mobile applications. Challenges 8. Available from: https://doi.org/10.3390/challe8010009.

Benos, L., Tagarakis, A.C., Dolias, G., Berruto, R., Kateris, D., Bochtis, D., 2021. Machine learning in agriculture: a comprehensive updated review. Sensors 21. Available from: https://doi.org/10.3390/S21113758.

Bir, C., Mintert, J.R., 2019. Farmer percepetions of precision agriculture technology benefits. J. Agric. Appl. Econ. 51, 142−163. Available from: https://doi.org/10.1017/aae.2018.27.

Campo, L.V., Ledezma, A., Corrales, J.C., 2020. Optimization of coverage mission for lightweight unmanned aerial vehicles applied in crop data acquisition. Expert Syst. Appl. 149, 113227. Available from: https://doi.org/10.1016/j.eswa.2020.113227.

Chivasa, W., Mutanga, O., Biradar, C., 2020. UAV-based multispectral phenotyping for disease resistance to accelerate crop improvement under changing climate conditions. Remote Sens. 12. Available from: https://doi.org/10.3390/rs12152445.

Cucci, D.A., Rehak, M., Skaloud, J., 2017. Bundle adjustment with raw inertial observations in UAV applications. ISPRS J. Photogramm. Remote Sens. 130, 1−12. Available from: https://doi.org/10.1016/j.isprsjprs.2017.050.008.

Dolias, G., Benos, L., Bochtis, D., 2022. On the routing of unmanned aerial vehicles (UAVs) in precision farming sampling missions. In: Bochtis, D.D., Sørensen, C.G., Fountas, S., Moysiadis, V., Pardalos, P.M. (Eds.), Information and Communication Technologies for Agriculture—Theme III: Decision. Springer International Publishing, Cham, pp. 95−124. ISBN 978-3-030-84152-2.

Dündar, Ö., Bilici, M., Ünler, T., 2020. Design and performance analyses of a fixed wing battery VTOL UAV. Eng. Sci. Technol. 23, 1182−1193. Available from: https://doi.org/10.1016/j.jestch.2020.020.002.

Elmeseiry, N., Alshaer, N., Ismail, T., 2021. A detailed survey and future directions of unmanned aerial vehicles (UAVs) with potential applications. Aerospace 8, 1−29. Available from: https://doi.org/10.3390/aerospace8120363.

Feng, A., Zhou, J., Vories, E.D., Sudduth, K.A., Zhang, M., 2020. Yield estimation in cotton using UAV-based multi-sensor imagery. Biosyst. Eng. 193, 101−114. Available from: https://doi.org/10.1016/j.biosystemseng.2020.020.014.

Gracia-Romero, A., Kefauver, S.C., Fernandez-Gallego, J.A., Vergara-Díaz, O., Nieto-Taladriz, M.T., Araus, J.L., 2019. UAV and ground image-based phenotyping: a proof of concept with durum wheat. Remote Sens. 11. Available from: https://doi.org/10.3390/rs11101244.

Guo, Y., Yin, G., Sun, H., Wang, H., Chen, S., Senthilnath, J., et al., 2020. Scaling effects on chlorophyll content estimations with RGB camera mounted on a UAV platform using machine-learning methods. Sensors 20, 1−22. Available from: https://doi.org/10.3390/s20185130.

Hardin, P., Jensen, R., 2011. Small-scale unmanned aerial vehicles in environmental remote sensing: challenges and opportunities. GISci. Remote Sens. 48, 99−111. Available from: https://doi.org/10.2747/1548-1603.48.1.99.

Hong, Y., Jung, S., Kim, S., Cha, J., 2021. Autonomous mission of multi-UAV for optimal area coverage. Sensors 21. Available from: https://doi.org/10.3390/s21072482.

Hunt, D., 1995. Farm Power and Machinery Management. Iowa State University Press.

Ivić, S., Andrejčuk, A., Družeta, S., 2019. Autonomous control for multi-agent non-uniform spraying. Appl. Soft Comput. J. 80, 742−760. Available from: https://doi.org/10.1016/j.asoc.2019.050.001.

Jackson, J., Ladino, K., Abdulai, G., Engineering, A., 2021 Decision Aid for Estimating the Cost of Using a Drone in Production Agriculture, Kentucky.

Jang, G.J., Kim, J., Yu, J.K., Kim, H.J., Kim, Y., Kim, D.W., et al., 2020. Review: cost-effective unmanned aerial vehicle (UAV) platform for field plant breeding application. Remote Sens. 12, 1−20. Available from: https://doi.org/10.3390/rs12060998.

Joint Undertaking, SESAR, 2016. European Drones Outlook Study: Unlocking the Value for Europe. Publications Office. Available from: https://www.sesarju.eu/sites/default/files/documents/reports/European_Drones_Outlook_Study_2016.pdf.

Ju, C., Son, H.Il, 2018. Multiple UAV systems for agricultural applications: control, implementation, and evaluation. Electron. 7, 1—19. Available from: https://doi.org/10.3390/electronics7090162.

Kim, J., Kim, S., Ju, C., Son, H.Il, 2019. Unmanned aerial vehicles in agriculture: a review of perspective of platform, control, and applications. IEEE Access 7, 105100—105115. Available from: https://doi.org/10.1109/ACCESS.2019.2932119.

Kim, S.B., Lee, S.H., 2020. Battery balancing algorithm for an agricultural drone using a state-of-charge-based fuzzy controller. Appl. Sci. 10. Available from: https://doi.org/10.3390/APP10155277.

Klauser, F., Pauschinger, D., 2021. Entrepreneurs of the air: sprayer drones as mediators of volumetric agriculture. J. Rural Stud. 84, 55—62. Available from: https://doi.org/10.1016/j.jrurstud.2021.020.016.

Lampridi, M.G., Kateris, D., Vasileiadis, G., Marinoudi, V., Pearson, S., Sørensen, C.G., et al., 2019a. A case-based economic assessment of robotics employment in precision arable farming. Agronomy 9, 175. Available from: https://doi.org/10.3390/agronomy9040175.

Lampridi, M.G., Kateris, D., Vasileiadis, G., Marinoudi, V., Pearson, S., Sørensen, C.G., et al., 2019b. A case-based economic assessment of robotics employment in precision arable farming. Agronomy 9, 175. Available from: https://doi.org/10.3390/agronomy9040175.

Lezoche, M., Panetto, H., Kacprzyk, J., Hernandez, J.E., Alemany Díaz, M.M.E., 2020. Agri-food 4.0: a survey of the supply chains and technologies for the future agriculture. Comput. Ind. 117, 103187. Available from: https://doi.org/10.1016/j.compind.2020.103187.

Liang, Y., Hong, C.Z., Yuan, Y., 2019. A review of rechargeable batteries for portable electronic devices. InfoMat. 1—27. Available from: https://doi.org/10.1002/inf2.12000.

Losè, L.T., Chiabrando, F., Tonolo, F.G., 2020. Boosting the timeliness of UAV large scale mapping. Direct georeferencing approaches: operational strategies and best practices. ISPRS Int. J. Geo-Inf. 9. Available from: https://doi.org/10.3390/ijgi9100578.

Martinez-Guanter, J., Agüera, P., Agüera, J., Pérez-Ruiz, M., 2020. Spray and economics assessment of a UAV-based ultra-low-volume application in olive and citrus orchards. Precis. Agric. 21, 226—243. Available from: https://doi.org/10.1007/s11119-019-09665-7.

Matese, A., Toscano, P., Di Gennaro, S.F., Genesio, L., Vaccari, F.P., Primicerio, J., et al., 2015. Intercomparison of UAV, aircraft and satellite remote sensing platforms for precision viticulture. Remote Sens. 7, 2971—2990. Available from: https://doi.org/10.3390/rs70302971.

Mattivi, P., Pappalardo, S.E., Nikolić, N., Mandolesi, L., Persichetti, A., De Marchi, M., et al., 2021. Can commercial low-cost drones and open-source gis technologies be suitable for semi-automatic weed mapping for smart farming? A case study in ne italy. Remote Sens. 13. Available from: https://doi.org/10.3390/rs13101869.

Mobarra, M., Rezkallah, M., Ilinca, A., 2022. Variable speed diesel generators: performance and characteristic comparison. Energies 15. Available from: https://doi.org/10.3390/en15020592.

Montes de Oca, A., Flores, G., 2021. The AgriQ: a low-cost unmanned aerial system for precision agriculture. Expert. Syst. Appl. 182. Available from: https://doi.org/10.1016/j.eswa.2021.115163.

Mueller-Stoffels, M., VanderMeer, J., Schaede, H., Miranda, L., Light, D., Simmons, A., et al., 2017. Diesel Generator Fuel Consumption Under Dynamic Loading.

Padró, J.C., Muñoz, F.J., Planas, J., Pons, X., 2019. Comparison of four UAV georeferencing methods for environmental monitoring purposes focusing on the combined use with airborne and satellite remote sensing platforms. Int. J. Appl. Earth Obs. Geoinf. 75, 130—140. Available from: https://doi.org/10.1016/j.jag.2018.100.018.

Petritoli, E., Leccese, F., Ciani, L., Roma, S., Navale, V., 2018. Reliability and maintenance analysis of unmanned aerial vehicles †. Sensors 18, 1−16. Available from:
https://doi.org/10.3390/s18093171.

Prior, S.D., Bell, J., 2011. Empirical measurements of small unmanned aerial vehicle co-
axial rotor systems empirical measurements of small unmanned aerial vehicle co-axial
rotor systems. J. Sci. Innov. 1, 1−18.

Pádua, L., Marques, P., Adão, T., Guimarães, N., Sousa, A., Peres, E., et al., 2019.
Vineyard variability analysis through UAV-based vigour maps to assess climate change
impacts. Agronomy 9. Available from: https://doi.org/10.3390/agronomy9100581.

Rasmussen, J., Ntakos, G., Nielsen, J., Svensgaard, J., Poulsen, R.N., Christensen, S.,
2016. Are vegetation indices derived from consumer-grade cameras mounted on
UAVs sufficiently reliable for assessing experimental plots? Eur. J. Agron. 74, 75−92.
Available from: https://doi.org/10.1016/j.eja.2015.110.026.

Reger, M., Bauerdick, J., Bernhardt, H., 2018. Drones in agriculture: current and future
legal status in Germany, the EU, the USA and Japan. Landtechnik . Available from:
https://doi.org/10.15150/lt.2018.3183.

Reynolds, D., Baret, F., Welcker, C., Bostrom, A., Ball, J., Cellini, F., et al., 2019. What
is cost-efficient phenotyping? Optimizing costs for different scenarios. Plant Sci. 282,
14−22. Available from: https://doi.org/10.1016/j.plantsci.2018.060.015.

Schirrmann, M., Giebel, A., Gleiniger, F., Pflanz, M., Lentschke, J., Dammer, K.H.,
2016. Monitoring agronomic parameters of winter wheat crops with low-cost UAV
imagery. Remote Sens. 8. Available from: https://doi.org/10.3390/rs8090706.

Shen, H., Zhang, Y., Mao, J., Yan, Z., Wu, L., 2021. Energy management of hybrid uav
based on reinforcement learning. Electron 10. Available from: https://doi.org/10.3390/
electronics10161929.

Sopegno, A., Calvo, A., Berruto, R., Busato, P., Bocthis, D., 2016. A web mobile application for agricultural machinery cost analysis. Comput. Electron. Agric. 130,
158−168. Available from: https://doi.org/10.1016/J.COMPAG.2016.080.017.

Späti, K., Huber, R., Finger, R., 2021. Benefits of increasing information accuracy in variable rate technologies. Ecol. Econ. 185. Available from: https://doi.org/10.1016/j.
ecolecon.2021.107047.

Subramaniam, R., Hajjaj, S.S.H., Gsangaya, K.R., Sultan, M.T.H., Mail, M.F., Hua, L.S., 2021.
Redesigning dispenser component to enhance performance crop-dusting agriculture drones.
Mater. Today Proc. Available from: https://doi.org/10.1016/j.matpr. 2021.030.015.

Sørensen, L.Y., Jacobsen, L.T., Hansen, J.P., 2017. Low cost and flexible UAV deployment
of sensors. Sensors 17, 1−13. Available from: https://doi.org/10.3390/s17010154.

Tamayo, L.V., Thron, C., Fendji, J.L.K.E., Thomas, S.K., Förster, A., 2020. Cost-
minimizing system design for surveillance of large, inaccessible agricultural areas using
drones of limited range. Sustainability 12, 1−25. Available from: https://doi.org/
10.3390/su12218878.

Tsiamis, N., Efthymiou, L., Tsagarakis, K.P., 2019. A comparative analysis of the legislation
evolution for drone use in oecd countries. Drones 3, 1−15. Available from: https://doi.
org/10.3390/drones3040075.

Vu, N.A., Dang, D.K., Le Dinh, T., 2019. Electric propulsion system sizing methodology
for an agriculture multicopter. Aerosp. Sci. Technol. 90, 314−326. Available from:
https://doi.org/10.1016/j.ast.2019.040.044.

Wang, J., Jia, R., Liang, J., She, C., Xu, Y.P., 2021. Evaluation of a small drone performance using fuel cell and battery; constraint and mission analyzes. Energy Rep. 7,
9108−9121. Available from: https://doi.org/10.1016/j.egyr.2021.110.225.

Wang, L., Huang, X., Li, W., Yan, K., Han, Y., Zhang, Y., et al., 2022. Progress in agricultural unmanned aerial vehicles (UAVs) applied in China and prospects for Poland.
Agriculture 12.

Xu, Y., Sun, Z., Xue, X., Gu, W., Peng, B., 2020. A hybrid algorithm based on MOSFLA and GA for multi-UAVs plant protection task assignment and sequencing optimization. Appl. Soft Comput. J. 96, 106623. Available from: https://doi.org/10.1016/j.asoc.2020.106623.

Yi, W., Sutrisna, M., Wang, H., 2021. Unmanned aerial vehicle based low carbon monitoring planning. Adv. Eng. Inform. 48, 101277. Available from: https://doi.org/10.1016/j.aei.2021.101277.

Zhang, J., Campbell, J.F., Sweeney II, D.C., Hupman, A.C., 2021. Energy consumption models for delivery drones: a comparison and assessment. Transp. Res. D Transp. Environ. 90, 102668. Available from: https://doi.org/10.1016/j.trd.2020.102668.

Zhao, X., Zhao, J., Wang, X., Ouyang, C., Li, R., Xiong, Y., et al., 2020. A procedure for power consumption estimation of multi-rotor unmanned aerial vehicle. J. Phys. Conf. Ser. 1509, 012015. Available from: https://doi.org/10.1088/1742-6596/1509/1/012015.

Zhu, H., Nie, H., Zhang, L., Wei, X., Zhang, M., 2020. Design and assessment of octo-copter drones with improved aerodynamic efficiency and performance. Aerosp. Sci. Technol. 106, 106206. Available from: https://doi.org/10.1016/j.ast.2020.106206.

CHAPTER 10

Energy efficiency for in-farm unmanned aerial system applications

Chris Cavalaris
Department of Agriculture Crop Production and Rural Development, University of Thessaly,
Volos, Greece

Introduction

Unmanned aerial systems (UASs), also commonly mentioned as drones, is a relatively new technology that is expanding rapidly with great potential in agriculture. Despite its wide acceptance and the high expectations, there are still some critical technical limitations to be resolved in order to allow effective and profitable field operations. Aerial platforms for instance should be able to work in large areas carrying the necessary payloads, provided a sufficient energy supply for the complement of their operation (Aslan et al., 2022). They should also be able to operate autonomously and reliably in a particular environment retaining constant communication with the base station.

Increased payload capacity is often a desirable feature for UASs in both crop monitoring where heavy type sensors, such as hyperspectral cameras and light detection and ranging (LIDAR) sensors, are sometimes necessary to monitor specific plant traits (Jang et al., 2020), as also during plant care product applications for a higher autonomy during a single sortie and thus, eliminating time losses for returning and refilling. A higher UAS overall weight also implies higher power demands and greater energy consumption. Flight-time duration therefore is shortened compromising the endurance of the UAS (P. Hardin and Jensen, 2011). A higher endurance implies a higher energy sink available onboard the UAS. Regardless of the utilized powering system, whether it is a battery, a fuel cell, or an engine using a fossil fuel, energy storage capacity is always associated to a carried mass. Therefore the approach of using larger energy sinks is compensated for the higher weight resulting again in increased power demands and reduced autonomy. Technological advances providing high density energy components, optimization of power use through intelligent energy management systems, optimum mission planning in swarms, or even, inflight recharging, are some promising solutions for improving economic feasibility of using UASs. However, it is the UAS type and payload that manifests primarily the flight performance and energy requirements.

Unmanned aerial system types — Energy aspects and performance

Agricultural UASs may be classified according to various criteria like their size, weight, configuration, motor type, power supply, range, endurance, and operational altitude (Elmeseiry et al., 2021). One common classification is according to their configuration. There are three prominent types: horizontal take-off and landing (HTOL), vertical take-off and landing (VTOL), and hybrid types.

Energy aspects of "Horizontal Take-Off and Landing" unmanned aerial systems type

HTOL UASs require a runway for accelerating during take-off and a runway for landing. Their aerodynamic design imposes a low engine thrust during flying offering limited energy requirements. They can carry heavy payloads and support long-range and endurance flights at high speeds. The most prominent type on this category are the fixed-wing aircrafts (Elmeseiry et al., 2021; Dündar et al., 2020). Fixed wing, however, cannot perform abrupt maneuvers and always need to move forward to remain airborne. They are most suitable therefore for high speed, long range, and endurance remote sensing applications at obstacle safe altitudes, allowing large areas to be covered in a limited amount of time, provided an adequate space for take-off and landing (Jang et al., 2020). Electric motor fixed wing for instance may provide an endurance of up to 4 h, while internal fuel combustion engines may reach up to 4.8 h (Santin et al., 2021). Unlike manned airplanes, however, they are not a preferable solution for plant care product applications because higher stall speeds are required when carrying high payloads. These high speeds at low altitudes impose serious safety issues in the absence of an onboard pilot.

Energy aspects of "Vertical Take-Off and Landing" unmanned aerial systems type

VTOL UASs do not require a runway as they have the ability to ascend and descend vertically. They present a high level of maneuverability and possess the ability to hover over points or areas of interest (Behjati et al., 2021). They are also less affected by weather conditions (Colomina and Molina, 2014). VTOL UASs therefore are most suitable for operations:

- in limited space or restricted areas,
- at low altitudes where obstacle avoidance might be an issue,
- that includes tasks at mountainous and complex rural terrain environments, and
- requiring stationary operations like spraying plant care products in trees.

The most prominent types are multirotor UASs comprising usually two, four, six, or eight rotor combinations. Single-rotor drones, another VTOL UAS configuration, are helicopter-type UASs with an ancillary tail rotor to encompass spin. VTOL UASs present poor aerodynamic performance since the horizontally mounted rotors impose an enormous increase in drag force, opposite to the direction of travel (Santin et al., 2021). Therefore they present higher power demands and increased energy consumption compared to fixed wing, reducing their endurance and capability for high range operations. The most common multirotors are electric powered providing an endurance of 20−30 min while more advanced models may reach a 75 min or up to 90 min autonomy (Elmeseiry et al., 2021; Boukoberine et al., 2019). Their mechanism also is much more complex than the one of the HTOL type, resulting on higher manufacturing cost (Elmeseiry et al., 2021). Multirotor agricultural UASs are gaining constantly advance over single rotor types because they are simpler and cheaper constructions, their operation is safer, and they possess lower noise and vibration (Ju and Son, 2018). Nevertheless, single rotors allow sufficient space to use large propellers that are able to produce a higher thrust force providing an increased payload capacity and an improved power efficiency (Santin et al., 2021). A recent evolution in multirotor UASs is utilizing dual coaxial rotors turning in opposite directions providing the opportunity to use also larger propellers without adding to the drone's overall size and allowing, therefore, a higher payload capacity (Jawad et al., 2022). One shortcoming is that the propeller at the lower level has a 22% thrust loss.

Energy aspects of "Hybrid" unmanned aerial systems type

Hybrid types of UASs is another configuration that is gaining increased interest. They are vertical take-off, fixed wing (VTOL-FX) drones that combine benefits of HTOLs and VTOLs. In fact, VTOL-FX are mainly fixed wing that utilize their propulsion to support VTOL (Dündar et al., 2020). After ascending to the operating altitude, they change their configuration into a fixed wing, reducing drag forces, power demands, and increasing endurance during their mission (Elmeseiry et al., 2021). There are different technical solutions implemented to achieve the transition, like tilting the rotors, tilting the wings, or tilting the whole fuselage. According to their function, they are classified into convertiplanes and tail-sitters (Saeed et al., 2015). Convertiplanes take off, cruise, hover, and land with their fuselage remaining always horizontal. Tail-sitters take off and land vertically on their tail while the fuselage tilts in a horizontal position during mission by utilizing differential thrust or control surfaces. The transition is made without the need of extra actuators, and therefore, their construction is lighter and simpler providing lower manufacturing cost and increased endurance. Nevertheless they are more vulnerable to the wind during the take-off and landing modes (Saeed et al., 2015).

Unmanned aerial systems power and energy demands

Drones must overcome gravity in order to lift and remain airborne, therefore they utilize a propulsion system that is powered by internal combustion or electric motors. The motors also provide power to overcome the body's and the rotors' air drag to be able to move into the airspace. Power is also required for the function of servo mechanisms and the drone internal electronics (Kirschstein, 2021). Sometimes, drones have to provide power to other onboard instruments like cameras and LIDARs. One of the main challenges in drone evolution is providing sufficient endurance and autonomy through adequate provision of power. The power, however, is generated by energy storage components carried upon the UAS that add weight to the UAS, which in turn, require extra power to overcome gravity. When larger gas tanks or batteries are used to increase flight duration, they impose an extra weight that reduce UAS autonomy (Cheng et al., 2020). It is a close loop cycle requiring alternative solutions to keep low the aircraft net weight and at the same time, increase endurance, and facilitate a higher payload capacity.

Some key factors affecting drone power demands and energy consumption are described in Vu et al. (2019) and are summarized in Table 10.1. Constant power provision is one of the main challenges in UASs technology for making their use efficient and cost-effective.

Table 10.1 Key factors affecting drone power demands and energy consumption.

Drone design	Environment	Drone dynamics	Delivery operations
• Drone weight • Number of rotors • Size of rotors • Size of drone body • Battery weight • Battery energy capacity • Size of battery • Power transfer efficiency • Maximum speed and payload • Lift-to-drag ratio • Delivery mechanism • Avionics	• Air density • Gravity • Wind conditions • Weather (e.g., rain and snow) • Ambient temperature • Regulations	• Airspeed • Motion (take-off, landing, hover, cruise) • Acceleration and deceleration • Angle of attack • Flight angle • Flight altitude • Riding on another vehicle	• Payload weight • Size of payload • Empty return • Fleet size and mix • Single-/multistop drone trip • Delivery mode (landing, tether, parachute) • Area of service region

Source: Adapted from Zhang, J., Campbell, J.F., Sweeney, D.C., Hupman, A.C., 2021. Energy consumption models for delivery drones: A comparison and assessment. Transp. Res. Part D Transp. Environ. 90, 102668. https://doi.org/10.1016/j.trd.2020.102668.

The limitations imposed by weight, size, and other aspects limit drone autonomy and involve considerable time for replenishing with energy. The problem is highly important for UASs with electric motors since fossil fuels powering UASs with internal combustion engines are of high energy density. There are many different technologies employed for providing electric power in drones such as batteries, fuel cells, hybrid systems, swapping, tethering, laser beam recharging, and others each having their pros and cons in power use efficiency and cost.

Unmanned aerial systems motors

There are two major types of motors used to power the propulsion system of the drones, internal combustion engines and electric motors. Accordingly, they use different kinds of power sources. Internal combustion engines use fossil fuels or/and biofuels while electric motors use electricity provided by batteries, supercapacitors, fuel cells, solar devices, and recently through wireless transmission. A third type is a hybrid kind of motor utilizing fuels and electricity for generating power.

Internal combustion engines

Internal combustion engines may use gasoline, methanol, or gas fuels. Gas engines, however, present good performances only for high power, above 100 hp ranges, therefore they are not suitable for small-scale UASs (Boukoberine et al., 2019). Gasoline and methanol are more common in small UASs and, thanks to their higher fuel energy and power densities (Hardin and Hardin, 2010), provide long endurance and large payload capacity despite the considerable energy losses imposed during the internal combustion cycle. Internal combustion engines require, however, an auxiliary electric motor to start and they are quite noisy during their operation (Boukoberine et al., 2019). Because of their mass, they increase the overall size and weight of the drone and impose operational risks due to less effective control by the remote operator (Erdelj et al., 2019). Their complexity increases the cost of the UAS, while the use of fossil fuels impose higher greenhouse gas emissions (Boukoberine et al., 2019).

Electric motors

Electric motors use electricity to produce mechanical power. There are two major types of electric motors used in drones: brushed and brushless. The brushless types are preferable because they present lower resistance providing higher efficiency. They are divided into outer-runner and interrunner according to the rotating part (array or magnet). Outer-runner offer low rotation speeds and high torque and are considered therefore as a better choice (Vu et al., 2019). Drones using electric motors are the most common because they present some substantial benefits like reliability, reduced energy losses through heat and high efficiency, low noise, zero direct pollutant emissions, self-starting, and rapid response enabling high maneuverability (Boukoberine et al., 2019).

Hybrid systems

Hybrid systems attempt to combine internal combustion engines with electric motors in a hybrid architecture in order to take advantage of the benefits of both types (Glassock et al., 2009). The system, however, appear to be highly complex with limited benefits in terms of fuel usage and endurance (Boukoberine et al., 2019).

Power supply systems

Batteries

A battery is an electrochemical energy pack able to convert chemical energy into electricity. Some batteries have also the ability to function vice versa, converting electrical energy into chemical energy while on charging mode, therefore, they are characterized rechargeable. Batteries may consist of a monolithic module, or a pack of several cells connected in a series or parallel mode. Their main features are their voltage (V), power (W), capacity (Wh), specific power (W/kg) power density (W/L), specific energy (Wh/kg), energy density (Wh/L), discharge rate, number of charge and recharge cycles, and operating temperature (°C), among others and other.

Batteries are the most common type of power source for small agricultural drones because they can easily adjust their power supply according to the changing power demands during a flight. They have a compact and simple design that can be easily mounted on the drone body, they are easily interchangeable, they are safe, and their cost is relatively low. Batteries, however, present relatively low specific energy and energy density that results in power shortages which limit the drone endurance. Using bigger or more batteries is a nonefficient option because the weight and the corresponding power demands are also increased. One of the biggest challenges for drone manufacturers, therefore, is the ability to use batteries with a higher energy density able to provide an extended power supply without compromising total weight.

Table 10.2 Key features of four common battery types used in drones.

Characteristic	Ni-Cd	Ni-Mh	Li-Po	Li-S
Specific energy (Wh/kg)	40	80	180	350
Energy density (Wh/L)	100	300	300	350
Specific power (W/kg)	300	900	2800	600

Source: Adapted from Boukoberine, M.N., Zhou, Z., Benbouzid, M., 2019. A critical review on unmanned aerial vehicles power supply and energy management: solutions, strategies, and prospects. Appl. Energy 255, 113823.

There are many different types of batteries that are utilized in drones providing different aspects of power supply. The most prominent are rechargeable types providing the chance of re-use over several lifecycles and therefore, reducing the cost of their use. According to their chemical composition, they are classified into different types, including lithium polymer (Li-Po), lithium ion (Li-Ion), nickel-cadmium (Ni-Cad), nickel-metal hydride (Ni-MH), and lithium iron phosphate (Li-Fe) (Table 10.2). Today, more than 90% of small drones use Li-Ion and Li-Po batteries that offer a good energy density at an affordable price (Vu et al., 2019). Their main difference is their electrolyte. Li-Po batteries use gel-like polymer electrolytes that aids in a more flexible form factor and lighter weight while Li-ion use liquid electrolytes based on lithium salts in organic solvents (Amanor-boadu and Guiseppi-elie, 2020). Li-ion batteries have several advantages, including a higher power and a high energy density, and low self-discharge rate (Amanor-boadu and Guiseppi-elie, 2020) being therefore, more suitable for lightweight drones. They have, however, a higher cost.

Another classification of batteries is based on their life cycle depending on the number charging and recharging cycles. Li-Po and Li-Ion batteries for instance may offer from 500 up to 2000 lifecycles (Dündar et al., 2020; Liang et al., 2019).

Fuel cells

Fuel cells are devices that convert chemical energy of a fuel (mainly hydrogen but also natural gas and biogas) and of an oxidizing agent (mainly oxygen) into electricity through a pair of redox reactions (Sarangi et al., 2018). Unlike batteries that carry a limited amount of stored energy, fuel cells may generate energy as long as fuel and an oxidizing agent are supplied to sustain the chemical reaction. Since the reaction agent is oxygen, that is regularly available in the air, the main constraint is the access to a hydrogen resource. There are different techniques employed currently in UASs to store hydrogen like compressed hydrogen gas, liquid hydrogen, and chemical hydrogen generation by reforming methanol or hydrocarbon fuels. Each of these storage techniques has its advantages and drawbacks, as discussed in Boukoberine et al. (2019).

Unlike internal combustion engines that convert chemical energy stored in fuels directly into mechanical power, fuel cells convert it initially into electricity and do not involve any combustion being though more efficient and quieter. Fuel cell efficiency is around 60% (Boukoberine et al., 2019), much higher compared to the under 30% for internal combustion engines, but also lower compared to the 90% or over for batteries (Boukoberine et al., 2019). Nevertheless, fuel cell is one of the most promising technologies for the future because they are a clean source of energy (the by-product is pure water) with a considerable high specific energy (Sarangi et al., 2018).

There are many fuel cell technologies employed for catalyzing the chemical reaction of the fuel and the oxidizing agent. They may be classified according to chemical criteria into catalysts, and electrolytes or according to their operation temperature into low, medium, and high. For UAS applications, the most suitable technology seems to be the polymer electrolyte membrane fuel cell (PEMFC), a low temperature fuel cell allowing less warm-up time and also some other essential characteristics for UASs such as lightweight, high power and energy density, long lifetime, and low response time to load variation (Boukoberine et al., 2019).

Due to their high specific energy, fuel cells offer increased endurance. LiPo batteries, for example, possess a specific energy up to 250 Wh/kg, while an equal weight fuel cell system can reach 1000 Wh/kg (Fig. 10.1). Fuel cell-powered drones can fly for hours instead of few minutes' endurance provided by batteries. In addition, they allow fast refueling contrary to the batteries that need a relatively long charging time (Boukoberine et al., 2021). The major constraint for utilizing the fuel cell technology in UASs is its currently high price which, however, is expected to become lower as more commercial applications will show up providing more feasible solutions (Suroso, 2018; MMC Unveils Next-Gen Hydrogen Fuel Cell for Use in Range of Drones, 2017).

Hybrid power systems

Besides the apparent profits, fuel cells have also some limitations when used as a unique power source for drones (Boukoberine et al., 2019).

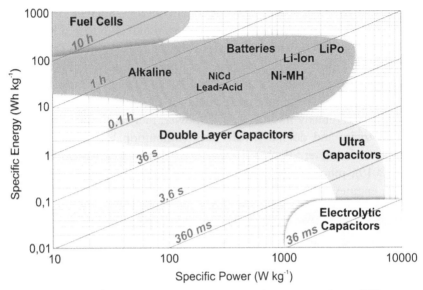

Figure 10.1 Specific energy/power comparison between various UAS energy sources. *Adapted from Boukoberine, M.N., Zhou, Z., Benbouzid, M., 2019. A critical review on unmanned aerial vehicles power supply and energy management: solutions, strategies, and prospects. Appl. Energy 255, 113823.*

Since they utilize pumps, valves, and compressors, they impose some mechanical flow delay at the order of some seconds resulting in inconsistencies in fuel supply and poor response to power peaks (Boukoberine et al., 2021). In turn, this may cause insufficient power provision compromising stability, reliability, and efficiency. To that point, a fuel cell power system supported by a battery pack or a supercapacitor, in a hybrid architecture may ensure a fast response to power demand variations and a higher overall efficiency especially when the drone performs typical maneuvers such as take-off and climbing. Supercapacitors are an alternative to batteries providing very fast energy storage, a much higher power density but also a much lower energy density, so their usage is ideal in hybrid systems for covering instantaneous power demands (Ruan et al., 2017). The fuel cell can be the major power system during cruising or descending while the supercapacitor or the battery is used as an energy buffer that can supply or even absorb power getting recharged by the fuel cell (Boukoberine et al., 2019). The use of dual or even multiple power supply system in UASs requires the implementation of an intelligent energy management system (EMS) regulating the power flows among the alternative power sources and the feeding of the propulsion system of the drone for optimizing the power use (Shen et al., 2021; Wang et al., 2020). In a recent study (Donateo and Spedicato, 2017), it was revealed that the performance and endurance of hybrid power systems in UASs were optimum when the mass of the system, including power sources EMS and other auxiliary devices, was higher than 7.3 kg. This limitation excludes mini low-cost drones from using such a system restricting the usage to bigger and more expensive devices.

Even though fuel cell technology is still under evolution, there are some commercially available hybrid products today that provide high endurance and performance solutions for drones. For instance, the Singapore-based HES Energy Systems developed an hybrid system using compressed gaseous hydrogen which was incorporated at a multirotor UAS and provided endurance for a 3.5 h flight plan (Adnan et al., n.d.). Another hybrid system presented by the Canadian EnergyOr Technologies was able to provide a flight plan of 10 h and 4 min (Corporation et al., 2011), while the South Korean MetaVista company was able to achieve a nearly 12 h and 7 min flight plan with a fuel cell power module adapted on a multicopter (Boukoberine et al., 2021). Nevertheless, these solutions are currently considered highly sophisticated, increasing considerably the cost of investment.

In-flight recharging

Besides fuel cell and hybrid systems, there are some other methods implemented in order to improve the drone endurance by delivering energy during the flight plans using various techniques like swapping, tethering, and wireless recharging.

- Swapping is a technique deploying ground stations for recharging the drone batteries during their mission (Boukoberine et al., 2019). The recharging may be performed by means of lead contacts, a laser beam or by replacing the depleted battery with another, fully charged without powering off the drone and without human intervention. The second case is called hot-swapping and the drone can continue immediately its mission (Santin et al., 2021). Since ground stations may include also a cost that is comparable to the cost of the drones, their placement should be optimized in order to reduce their number and enhance the feasibility of the technique (Tamayo et al., 2020). The aim is to determine the path of the drone that minimizes the number of locations for recharging stations as well as the completion time of the mission (Fendji et al., 2020). There are various approaches for properly allocating the ground stations depending on the flight plans, the power demands, and the accessibility of the region. Recharging stations can even be small unmanned ground vehicles (UGVs) moving to the next appropriate point after each recharging. This approach allows a cost-effective path planning on large farms. The swapping technique is suitable only for VTOL drones that use computer vision to identify the exact place of the recharging station and land over it.

- Tethering is a technique for powering the UASs via a wire connecting permanently the drone with a ground-based power supply (Boukoberine et al., 2019). That way it is provided unlimited autonomy, but the main constraint is the limited distance that the drone can move away from its base station. There is no need to carry any batteries, but weight may be added through the cable especially if it is made from copper. Fiber cables are much lighter, taking place though in tethered drone innovations but their technology is much more expensive. Tethering is suitable for obstacle free, small areas for intensive operations that are, however, rarely needed in agriculture.
- Wireless recharging was proposed as an alternative approach to tethering. A laser generator placed on a ground power station transmits a light beam to the UAS while it is airborne while an embedded optical receiver converts the light to electricity for powering the UAS (Boukoberine et al., 2019). For that scope, the drone approaches a special aerial power link area without need for landing. There is also a chance to use UGVs moving to appropriate positions for effectively delivering the laser beam to the drone.

Conclusions

Power source endurance providing high flight autonomy is of the most critical challenges for making UASs efficient and cost-effective and allowing their wide scale adoption in agriculture. Increased payloads to carry pesticides, fertilizers, and other agricultural supplies as also various sensors impose the need of an onboard power resource with a high energy density and reduced size and weight. Electric-based sources seem to be more suitable for powering small agricultural UASs since they provide simplicity and high flexibility in energy management compared to traditional internal combustion engines that are quite complicated constructions with limitations on adjusting their power supply to the demands. The major constraint in electric power resources like the Li-Po and Li-Ion batteries that are mainly used today in UASs is their relatively low energy density that limit the load capacity and flight endurance. Fuel cells using hydrogen provide the opportunity for onboard generation of electricity being one of the most promising future technologies for long time UASs powering. Hybrid-powered systems, swapping, tethering, laser beam recharging, and others have been also proposed as efficient and low-cost solutions, each one imposing its own constraints. Optimization of power use through intelligent energy management systems, optimum mission planning, and other approaches can further improve efficiency. Finally, different types of UASs like the HTOL, the VTOL, and the hybrid types impose different characteristics of energy usage, hovering, and maneuverability and may be more suitable and effective for alternative tasks.

References

Adnan, N. et al., n.d. An overview of electric vehicle technology. In: Intelligent Transportation and Planning, IGI Global, 292–309.

Amanor-boadu, J.M., Guiseppi-elie, A., 2020. Improved performance of Li-ion polymer batteries through improved pulse charging algorithm. Appl. Sci. (Switz.) 10 (895).

Aslan, M.F., et al., 2022. A comprehensive survey of the recent studies with UAV for precision agriculture in open fields and greenhouses. Appl. Sci. (Switz.) 12 (3).

Behjati, M., et al., 2021. Lora communications as an enabler for internet of drones towards large-scale livestock monitoring in rural farms. Sensors 21 (15).

Boukoberine, M.N., et al., 2021. Hybrid fuel cell powered drones energy management strategy improvement and hydrogen saving using real flight test data. Energy Convers. Manag. 236, 113987.

Boukoberine, M.N., Zhou, Z., Benbouzid, M., 2019. A critical review on unmanned aerial vehicles power supply and energy management: solutions, strategies, and prospects. Appl. Energy 255, 113823.

Cheng, C., Adulyasak, Y., Rousseau, L.M., 2020. Drone routing with energy function: formulation and exact algorithm. Transport. Res. B 139, 364–387.

Colomina, I., Molina, P., 2014. Unmanned aerial systems for photogrammetry and remote sensing: a review. ISPRS J. Photogramm. Remote Sens. 92, 79–97.

Corporation, C.E., Bremen, N., Frontier, O.T., 2011. Energy or fuel cell powered UAV reaches 10h flight endurance. Fuel Cell Bull. 2011 (9), 4–5.

Donateo, T., Spedicato, L., 2017. Fuel economy of hybrid electric flight. Appl. Energy 206, 723–738.

Dündar, Ö., Bilici, M., Ünler, T., 2020. Design and performance analyses of a fixed wing battery VTOL UAV. Eng. Sci. Technol. 23 (5), 1182–1193.

Elmeseiry, N., Alshaer, N., Ismail, T., 2021. A detailed survey and future directions of unmanned aerial vehicles (UAVs) with potential applications. Aerospace 8 (12), 1–29.

Erdelj, M., Saif, O., Natalizio, E., Fantoni, I., 2019. UAVs that fly forever : uninterrupted structural inspection through automatic UAV replacement. Ad Hoc Netw. 94.

Fendji, J.L.E.K., et al., 2020. Cost-effective placement of recharging stations in drone path planning for surveillance missions on large farms. Symmetry 12 (10), 1–25.

Glassock, R., Hung, J., Gonzalez, L., Walker, R., 2009. Multimodal hybrid powerplant for unmanned aerial systems (UAS) robotics. In: Twenty-Fourth Bristol International Unmanned Air Vehicle Systems Conference, March 30th to April 1st 2009, Bristol United Kingdom.

Hardin, P.J., Hardin, T.J., 2010. Small-scale remotely piloted vehicles in environmental research. Geogr. Compass 4 (9), 1297−1311.

Hardin, P., Jensen, R., 2011. Small-scale unmanned aerial vehicles in environmental remote sensing: challenges and opportunities. GISci. Remote Sens. 48 (1), 99−111.

Jang, G.J., et al., 2020. Review: cost-effective unmanned aerial vehicle (UAV) platform for field plant breeding application. Remote Sens. 12 (6), 1−20.

Jawad, A.M., et al., 2022. Wireless drone charging station using class-E power amplifier in vertical alignment and lateral misalignment conditions. Energies 15 (4).

Ju, C., Son, H.I., 2018. Multiple UAV systems for agricultural applications: control, implementation, and evaluation. Electronics (Switz.) 7 (9), 1−19.

Kirschstein, T., 2021. Energy demand of parcel delivery services with a mixed fleet of electric vehicles. Clean. Eng. Technol. 5, 100322.

Liang, Y., Hong, C.Z.Z., Yuan, Y., 2019. A review of rechargeable batteries for portable electronic devices. InfoMat 1−27.

MMC unveils next-gen hydrogen fuel cell for use in range of drones. 2017. Fuel Cells Bull. 2017(8): 5.

Ruan, J., Walker, P.D., Zhang, N., Wu, J., 2017. An investigation of hybrid energy storage system in multi-speed electric vehicle. Energy 140, 291−306.

Saeed, A.S., Younes, A.B., Islam, S., Dias, J., Seneviratne, L., Cai, G. 2015. A Review on the Platform Design, Dynamic Modeling and Control of Hybrid UAVs. International Conference on Unmanned Aircraft Systems (ICUAS), Denver Marriott Tech Center, Denver, Colorado, USA, June 9-12, 2015

Santin, R., Assis, L., Vivas, A., Luciano, C.A.P., 2021. Matheuristics for multi-UAV routing and recharge station location for complete area coverage. Sensors 21 (5), 1−34.

Sarangi, P.K., Nanda, S., Mohanty, P., 2018. Recent advancements in biofuels and bioenergy utilization.

Shen, H., et al., 2021. Energy management of hybrid UAV based on reinforcement learning. Electronics (Switz.) 10 (16).

Suroso, I., 2018. Compressed hydrogen storage in contemporary fuel cell propulsion systems of small drones compressed hydrogen storage in contemporary fuel cell propulsion systems of small drones. In: IOP Conf. Series: Materials Science and Engineering 421 (2018) 042013 IOP.

Tamayo, L.V., et al., 2020. Cost-minimizing system design for surveillance of large, inaccessible agricultural areas using drones of limited range. Sustainability (Switzerland) 12 (21), 1−25.

Vu, N.A., Dang, D.K., Le Dinh, T., 2019. Electric propulsion system sizing methodology for an agriculture multicopter. Aerosp. Sci. Technol. 90, 314−326.

Wang, M., Shuguang, Z., Johannes, D., Florian, H., 2020. Battery package design optimization for small electric aircraft. Chin. J. Aeronaut. 33 (11), 2864−2876.

Zhang, J., Campbell, J.F., Sweeney, D.C., Hupman, A.C., 2021. Energy consumption models for delivery drones: A comparison and assessment. Transp. Res. Part D Transp. Environ. 90, 102668. Available from: https://doi.org/10.1016/j.trd.2020.102668.

CHAPTER 11

A gender perspective on drones and digital twins in a transdisciplinary European smart agriculture and forestry project

Wendy Wuyts and Lizhen Huang
Department of Manufacturing and Civil Engineering, Norwegian University of Science and Technology, Gjøvik, Norway

Introduction

Recently, the European Union and other funders demanded a more responsible research and innovation approach in the projects, as a reaction to the increasing evidence of research and design biases that led to the exclusion and even discrimination of certain population groups (Schiebinger, 2021). Research needs to be rooted in the experiences and needs of different groups of citizens. However, down the centuries, educated elites have shaped and dominated research and defined knowledge and learning, which led to some groups of citizens being overlooked. Hence, the integration of a gender dimension is one of the requirements in many Horizon Europe calls. This is based on the emerging field of research on gendered innovations, which

> *employs methods of sex, gender, and intersectional analysis to overcome past bias and, importantly, to create new knowledge. It seeks to harness the creative power of sex, gender, and intersectional analysis for innovation and discovery. The operative question is: does considering these factors add valuable dimensions to research? Do they take research in new Directions?*
>
> ***(Schiebinger, 2021).***

The emerging subfield of gendered innovations did not intersect with all fields yet, and there are a lot of underexplored frontiers (Schiebinger, 2021). Although there is a whole body of research on gender analysis in rural areas and innovation projects that show how they often not include women (e.g., Petesch et al., 2018), useful frameworks were not found with practical straight-to-the-point guidelines and steps that can help

Unmanned Aerial Systems in Agriculture
DOI: https://doi.org/10.1016/B978-0-323-91940-1.00011-6

engineers and solution providers, to identify and understand the gender dimension in multidisciplinary forestry and agriculture-related projects where technologies such as digital twins (DTs) and drones are central, which can be unfolded in many subtasks from standardization and reference models on trustworthiness via algorithms in visual recognition to gender analysis of impacts related to animal welfare.

This chapter provides a perspective as the fruit of the first phase of SPADE — multipurpose physical-cyber agri-forest drones ecosystem for governance and environmental observation, a research and innovation project (2022—2026) that includes tasks on the gender dimension and a work package on ethics (European Commission, 2022). Although the emerging subfield provided evidence mostly based on historical analysis, we did not find supporting guidelines and frameworks for identifying biases and opportunities for innovation and knowledge creation in solutions that are not yet developed and deployed in the real world. There are a multitude of guidance documents, mostly from the EU (e.g., European Union, 2020). However, there are not many reports on how this has been implemented in practice (e.g., Søraa et al., 2020). Despite operational rules, guiding questions, and gender dimension plans for domains such as health and machine translation (e.g., World Health Organization, 2002; European Union, 2020), historical evidence on gendered innovations in smart forestry and agriculture, as well as frameworks for future designs and research, is not coherent. This chapter provides a short overview of definitions, operational rules, and recommendations that are relevant for the Horizon Europe project SPADE, tailored for drone and DT engineers, operators, and researchers in smart forestry and agriculture.

Impact-driven projects require knowledge from different disciplines and domains, and across different contexts, so gender dimension tasks are a horizontal component that intersects all these fields and requires not only a reflection on each identified task and subtask in such a research project but also a systems thinking approach and looking for relationships and interconnections between gender dimensions of the different tasks that negatively or positively influence each other. This leads to the collection of fragmented information. As a preparatory action, an explorative critical literature review (Snyder, 2019) in different fields that intersect smart forestry, agriculture, and rural impacts was taken place, and critical questions — by email and short virtual roundtables — have been asked to various domain experts within the Horizon Europe project SPADE, with the mission to make a framework and a list of practical steps for different research and innovation tasks. The new perspective builds on previous social science research on gender dimensions in rural revitalization and innovation projects. It also includes inputs from domain experts in both DTs and drones in smart agriculture and forestry. This is significant as they read things differently than experts in social science and humanities, so authors had to tweak the outputs of the exploratory phase (guidelines and frameworks) so it works for experts in all these fields.

Gender dimension

First, the gender dimension encompasses more than just gender roles. There is a difference between gender and biological sex. The most cited scholars are West and Zimmerman who describe gender as *"the activity of managing situated conduct in light of normative conceptions of attitudes and activities appropriate for one's sex category"* (1987:127). Sex category is *"a categorisation established and sustained by the socially required identificatory displays that proclaim one's membership in one or the other category"*. (1987:127).

West and Zimmerman argue that gender is not something we are, but something we do. "Doing gender involves a complex of socially guided perceptual, interactional, and micropolitical activities that cast particular pursuits as expressions of masculine and feminine 'natures'" (1987:126).

In other words, gender is expected, a product of social doing, and often seen as a natural given, but it is created by society, and people are often not aware of that.

In fact, people often mistakenly believe that the gender dimension is about women. First of all, if women thrive, this will have consequences for the men around them. Second, different gender analysis studies also focused on a specific group of men (mostly low-income men) who can also be *losers* when certain innovations and transitions occur, because of structural reasons that have less to do with the individual but more with expectations, rules, and other social structures of society in this context.

In addition, the motivation behind gender draws from the international body of literature on women's social associations with the natural environment (Ortner, 1972; Plumwood, 1993). *"Women-environment connections — especially in reproductive and subsistence activities such as collecting fuel wood, hauling water, and cultivating food — were often presented as if natural and universal rather than as the product of particular social and cultural norms and expectations"* (UN Women, 2014). As Ortner (1972) noted, this relationship between nature and women is part of gender expectations, which are only present in certain cultures at a certain time. Additionally, Judith Butler (1990) wrote that gender is a product of culture and history. In certain cultures, mostly in rural areas in developing countries, the relationship between gender roles and the environment is still visible. This is because women and men live in separate spheres, and women with their closeness to nature have different perceptions and knowledge than men. The hypothesis underlying some "gender dimension" analyses concerns gender discrimination more than sex discrimination. Here are four possible hypotheses to illustrate the difference:

1. Women would benefit from assistance from drones and DTs, because their time in the field would be reduced, and they could combine remote monitoring work more easily with their care work.

    This has to do with gender. Generally speaking, women do more care work, especially in more traditional and conservative rural areas, and they are more inclined to stay at home. DTs and remote monitoring can lead to more chances to work from home. This allows more in-office or home-based work using digital technologies and calculations. Moreover, these tasks are rather based on the brain than the brawn (Wajcman, 2010). In addition to the fact that this can be related to gender expectations that women should not do the same physical work as men, remote monitoring can be useful for elderly people and anyone who is unable to move around easily.

2. Women would not likely adopt the DT solutions, because of the perception that forestry and working with digital technologies like drones and knowledge management are a "man's job." It is again related to the gender dimension and the expectations of society on gender roles. This has nothing to do with biological sex.

3. Drone operation is more difficult for women because the design is based on male bodies.

Biological sex plays a significant role in engineering and product design research. Sex includes anatomical and physiological characteristics that may affect **the design of products, systems, and processes** (see Schiebinger et al., 2011-2021). Many devices and machines have been designed to fit male bodies. For example, military and commercial cockpits were traditionally based on male anthropometry, which made it difficult or even dangerous for some women (or small men) to be pilots (Weber, 1997). Office building thermostats, which are based on male metabolic rates, may set temperatures too low for many women (van Hoof, 2015). In drone technology, also body poses and ergonomics are affected and affect the user experience, especially when commanding drones through body poses (Yam-Viramontes et al., 2022). Similarly, it is crucial to determine how drones react to non-White or female bodies in wheelchairs, for example.

Standards lead only to positive impacts for men, not women

This again has to do with biological sex and is a consequence of the previous statement. A large study by SCC revealed that "countries which are more involved in standardization experience fewer unintentional male deaths. As a country's participation in standardization increases, the number of men who die as a result of unintentional injuries decreases. When the analysis was repeated to determine the impact of participation in standardization on the number of unintentional female deaths, there was no impact. Unlike for men, increasing participation in standardization is not associated with a decline in the number of women who die as a result of unintentional injuries. Standards are not protecting women as well as they protect men" (SCC, 2020). Therefore, there are initiatives in standardization bodies like ISO, to investigate the impact of standards on women and adapt them.

From gender dimension to an intersectionality analysis

Gender dimensions in funding calls entail the risk that they can lead to a tunnel vision of gender (and/or biological sex), especially if there are no researchers involved who do not know intersectionality theories. It is not enough to look at gender alone, which is quite limiting and even discriminatory and excluding. By referring to "gender issues," the EU works against the cultural and societal need that all citizens be treated equally under all European legislative norms, regulations, and frameworks. Talking about gendered innovations and dimensions is pretty 1970s, pre-Kimberlé Crenshaw's famous essay on the need for intersectionality analysis (Crenshaw, 1989). A Black woman, for example, may be treated differently than White women and especially White men when applying for microcredits to buy a digital solution because of racial or gender-based discrimination. Hence, it is also imperative to understand differences within groups of women, men, and gender-diverse people. We have to look at everyone and respect everyone's needs for information and limits.

Therefore, we should also deploy an intersectional analysis and look at combinations of different social categories, for example, gender and age, or gender and access to networks (internet and social networks).

However, there are studies that tend to focus on the influence of one variable, and not combinations. This can lead to conflicting ideas. For example, regarding social acceptance of drones, we saw how gender can play a role (e.g., Wachenheim et al., 2021), but age was not significant. In another paper more focused on drone operation and familiarity with drone-related terminology research, both age and gender were identified as significant demographic aspects within the surveys (Sabino et al., 2022). But again, it was not clear if the combination mattered. In SPADE, and from this perspective, we focus not only on sex and gender, but we also consider intersectional factors, such as race and ethnicity, socioeconomic status, age, abilities, geographical location, family structure, and access to networks in our case studies, inspired by previous research (e.g., Elias et al., 2018).

Never forget the context (and the other structural problems in the deployment areas)

First of all, gender roles and how they influence social acceptance of solutions can be different within Europe, because some deployment areas are more conservative than others.

Secondly, sometimes projects with positive intentions can lead to negative impacts if we do not look at other problems in the deployment areas of these solutions. Remote monitoring can reduce on-site work and increase more "home-based work," but as forestry and farming are often considered male jobs, the hypothesis is that if men are more at home, the likelihood of domestic and gender-driven violence can increase, especially given that these men work in highly demanding sectors with low margins and if there are no other meaningful jobs in this same area. On the other hand, technologies that enable women to work in more safe environments (e.g., home-based offices) lower the likelihood of gender-based violence at work sites and in fieldwork. Although the numbers are comparatively low, some women might actually avoid home-based working situations (as a result of the increasing likelihood of domestic violence) and even sabotage the implementation of technologies that reduce time away from home. Therefore, the EU recommends — for projects dealing with rural areas — *"to investigate and include these specific risks as well as prevention and resilience options"* (European Union, 2020).

The SPADE project (2022–2026)

The strategic objective of SPADE is to develop an intelligent ecosystem to address the multiple purposes concept in light of deploying unmanned aerial vehicles (UAVs) to promote sustainable digital services for the benefit of a large scope of various end users in the sectors of forestry, cropping, and livestock farming. This includes individual UAV usability, UAV type applicability (e.g., swarm, collaborative, autonomous, and tethered), UAV governance models availability, and trustworthiness. Sensor dataspace reusability will be determined based on trained artificial intelligence and machine learning models. One of SPADE's departure points is that these models will enable sustainability and resilience in the overall life cycle of developing, setting up, offering, providing, testing, validating, and refining, as well as enhancing digital transformations and "innovation building" services in agriculture (European Commission, 2022).

First, SPADE will create a digital platform that will be able to realize the potential benefits to be reaped from the use of drones. This platform makes drone operations more accessible and controllable and also provides a service channel for value-added services enabled by drones.

Second, SPADE illustrates three innovative use cases for drones by making use of the digital platform. While demonstrating the use cases, the benefits coming from the use of drones are analyzed and quantified on a detailed stakeholder-level basis. This will demonstrate the new business opportunities. Demonstrations/pilots will also serve as a platform to study the regulatory framework at a national and international level. Three prototypes of drones are allocated to the domains of forestry, cropping, and livestock farming. This means deploying nine test cases. Additionally, a common DT framework for the management of heterogeneous agricultural and environmental applications and better interoperability for the digital solutions will be developed.

Finally, the orchestration of drone operations for heterogeneous use cases through the DT representations of agricultural and environmental assets will also be provided, thus completely changing the paradigm on how this drone equipment is used and benefited in practical applications (SPADE Website, 2023).

We identify SPADE as a smart farming and forestry project because it uses sensor and monitoring technologies to support forestry and farming practices while analyzing data from these practices in order to guide foresters and farmers in making sustainable decisions with less risks

(Vate-U-Lan et al., 2016; Krishna, 2017). The digitalization of farming and forestry has dramatically changed what it means to be a farmer or forester. There are new opportunities, but also new risks, related to privacy issues and surveillance (Klauser, 2018). Rural development will lead to unequal access to, adoption of, and use of ICT in rural areas. In this digitalization trend, men and women with fewer digital skills will be at a loss, and this is equally true for SPADE. Since women are often encouraged to avoid ICT education, there is a hypothesis that projects such as SPADE might lead to more discrimination against women. But not only women but also men, and especially men without digital skills, will also be victims. Rather than pushing them to be all ICT gurus, a gender dimension will require user interfaces to have a low threshold.

Methodology

In the agricultural sector, academic literature addressing gender, age, and other social attributes is fragmented and limited. Recent literature rooted in Asia and Africa has discussed the adoption of pesticides in China (Wachenheim et al., 2021). Integrating a gender dimension and identifying and implementing measures would not only benefit individuals but also society and the (rural) agriculture and forestry sector in general. Research about the agricultural sector shows that gender and other normative social structures can lead to unequal benefits from innovation opportunities (Petesch et al., 2017). As preparation, we did a critical literature review, because the sample of academic literature on the interaction of smart agriculture, drones, and DT with the gender dimension was very limited. Search words were smart, drone, and DT — in combination with gender, men, women, or ethics. There are no exclusion criteria, but it is limited to English literature. The key words were inspired by previous research and observations of rural revitalization projects in Europe and Asia.

We included reports (e.g., the EU report (2020) on gendered innovations and findings and guidelines of the Consultative Group on International Agricultural Research (CGIAR) on enabling gender equality in agricultural and environmental agriculture). The CGIAR created enabling gender equality in agricultural and environmental innovation (Gennovate), aimed at developing gender-transformative approaches that embed gender norms in innovation processes. We have read papers on gender, technology, and innovation in domain-specific journals, as well as blogs and deliverables, of national and European projects. An invaluable source is the website Gendered Innovations, initiated by Stanford University in 2010, and the research group behind was involved in the inclusion and evaluation of gender dimension in European Commission programs as Horizon 2020 and Horizon Europe (Schiebinger, 2021). As one of the key topics in SPADE is standardization, we looked into ISO standards (e.g., ISO, 2022) and analyzed their proposed steps for "women in standards," as well as the argumentation for more gender-responsive standards.

We developed a framework where we indicated which domains (digital innovation, with a separate column for drones, agriculture, or both) the literature addressed. We read all the materials through predefined categories distilled from literature on procedural, distributional, and recognition-based justice in governance, inspired by social research on justice in energy transitions (Jenkins et al., 2016). While there are many unexplored frontiers, the critical reading was mostly abductive since some digital innovations (e.g., data twins, drones, and machine learning) are rapidly developing and changing the landscape. One danger in this critical literature review approach (Snyder, 2019) lies in "cherry picking," and therefore this draft was validated by different domain experts (drone development, DT development, and agriculture experts) within the Horizon Europe SPADE project and external peers who are more specialized in gender and digital innovation. They were asked to validate and pinpoint systemic gaps.

In a transdisciplinary project like the SPADE project, the gender dimension requires knowledge of possible impacts and awareness of ethical guidelines for multiple disciplines and domains, from visual recognition algorithms to animal behavior. It is not enough to have only women or female testers/users in the consortium but to have consortium members trained in gender analysis (Schiebinger et al., 2011—2021). Moreover, the interdisciplinary nature can lead to knowledge and insights getting lost in translation. For example, ICT engineers use other jargon and work with different mental models than, for example, a farmer. Therefore, we developed a simple matrix to facilitate navigation through all the various disciplines and domains involved. In transdisciplinary research, a simple matrix can help in navigating different epistemologies and gender dimensions (Nikulina et al., 2019).

After desktop-based research in the first months of the SPADE project, we developed a Miro board that echoes a matrix of columns and rows for each work part of the project (Fig. 11.1).

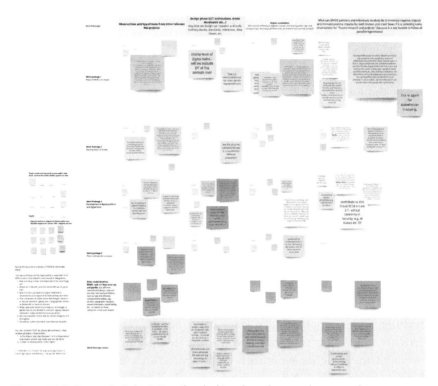

Figure 11.1 Screenshot depicting the desktop-based research approach.

The first columns are inspirational; they are findings from gendered innovations, often in other domains, that can help signal/identify biases in the design or hypotheses for the impact aspect. In the second column, we propose ideas about biases and other concerns in the design phase, which can relate to algorithms (e.g., visual recognition). The third column collects hypotheses about direct and indirect impacts. The fourth column collects inputs on steps that the SPADE project can take in the next years of research and development. During the meetings, we made notes and added more post-its to the Miro board. Before we shared the Miro board link, we clarified some basic ideas, e.g., that gender is not only a concern for women but also for men. We discussed the fact that farmers and foresters can be low-income men and that our DTs and drones can benefit them in terms of income, but also safe working conditions. We invited consortium members to visit the Miro board, add post-its, and give comments on the board or by email. This led to further reflections and negotiations on the framework and scope of the SPADE project itself. We acknowledge that we cannot study each hypothesis given the limits of time and budget. In the same period, we finalized the first version of the ethical guidelines, which led to a cross fertilization of these reflections.

As the SPADE project wanted to develop a common DT framework, we did not split up into further domains (livestock, crops, and forestry), but we are still aware that certain use cases involve animals and therefore a biological sex and gender dimension. We had to ask livestock experts if a ewe could react differently to a drone than a ram. We used different colored post-its for different categories, such as negative and positive impacts and female and male bodies (biological sex). However, there are limitations, and we used other colors if the impact was ambiguous or if it had intersectionality with other social categories (age, digital literacy, access to horizontal networks, etc.).

Recommendations and arguments

Have gender balance in your research and development team

The first low-hanging fruit, although not so easy for industries like forestry, farming, and ICT sectors that are shaped by gender stereotypes, is to have enough women in an R&I team, and not only in the roles of secretary, administration, management, and communication. This includes the promotion of the positions of women, girls, and marginalized groups. There is a structural problem of not having enough women in higher positions in R&I and in agriculture and forestry, and often the very few women in these sectors are overburdened/overasked with requests to attend boards and projects just because of their biological sex and minimum gender quota. During the SPADE proposal-writing/negotiation process, the men expressed a very open and embracing response to an entire chapter dedicated to gender impact, contributing their concerns and ideas to this chapter.

However, gender balance is not enough. Being a woman does not mean you are trained in gender analysis. Therefore, we need recommendations, ethical guidelines, and a framework that can guide all involved researchers, designers, and other innovation actors in SPADE and relevant projects. Designing sex, gender, and intersectional analysis into research is one crucial component contributing to world-class science and technology (Tannenbaum et al., 2019).

Avoid negative impacts through design
Embrace participatory design and involve the unrecognized agents
In development and design research, there are different levels of engagement with the potential users of the solutions. One of the main recommendations is to employ cocreation and participatory research that include end users when developing improved technologies, services, and business models (Macinnis-Ng and Zhao, 2022), and preferably on multilevels (level of users, like a farmer and forester, level of intermediaries, and level of governments). The choice of users and participants in testing and codeveloping solutions should include attention to gender, age, and socioeconomic status so that solutions appeal to users in their individual contexts (European Union, 2020). Participatory methods are widely embraced, but it is important to also be conscious of how events construct "local knowledge" in ways that are strongly influenced by existing social relationships. It suggests that information for planning is shaped by relations of power and gender, and by the investigators themselves, and that certain kinds of knowledge are often excluded (Mosse, 1994). There are risks that women and the elderly are considered the "other" in designs, which leads to excluding designs (Oudshoorn et al., 2016).

Therefore, the participatory design approach should make extra efforts to locate the invisible: dropouts and vulnerable groups of possible end users. If they are identified, developers can start listening (instead of proposing solutions) and sitting at the table with crop farmers and foresters in Spain, Greece, Norway, and other use cases that will test the SPADE solution. Designers can ask: Why do you (not) use drones, and for what applications? This helps us understand their personal requirements. It is also advised to record their concerns. As a result, business and governance model developers can propose benefit-sharing, risk-sharing, and profit-distribution models across gender and other social categories. A hypothesis is that we would have different use cases for drones, sensors, and DTs if women active in rural areas were interviewed, because they have different complaints and needs (see, e.g., Balearic Islands). For the Norwegian pilot, there is a larger difference between women and men in this regard, where women forest owners more strongly value the ecological, recreational, and social aspects of their forests (Nordlund & Westin, 2011; Umaerus et al., 2019), even if men and women are equally interested in timber production. Women forest owners are also more interested in forest-related businesses (Laszlo Ambjörnsson, 2021). This could also lead to different results. Secondly, when we organize events for transdisciplinary knowledge creation, we need to have trained facilitators who can navigate epistemological, gender, and linguistic diversities without discriminating against any source of knowledge (Nikulina et al., 2019).

There are also ethical concerns around how knowledge is created in and with DTs (Van Der Burg et al., 2021). One may question; who asked for the twin and whose interests it serves. Are decisions about what the twin should represent made solely by the developers of the DT (e.g., the scientists), which means it is their ideas, values, and assumptions about the issue that are included, or does a larger group of stakeholders have a say in that? And do the "big data" on which DTs are based introduce a bias by themselves, given that they do not represent everything (e.g., the entire animal population or the full range of possible scenarios), and they may likely lack or have insufficient data on small farms or rare phenomena? Do we only focus on the needs and interests of big farms? (Van Der Burg et al., 2021; Manfre et al., 2013).

As SPADE is not in the data collection phase, or even in reference model development that includes data collection, this will have to be seen.

In some cases, not all agents can be interviewed. In SPADE, animal welfare is an ethical concern. We cannot ask goats or sheep what they want. However, we should acknowledge their intelligence. As a result of goats learning about GPS shadows, they outwitted the GPS solutions in the Norwegian Nofence project (where fences are replaced by a band around the neck of the goats) (Søraa and Vik, 2021). Plants have intelligence too and will adapt their behavior. There is evidence that the sound of water influences the behavior of bean plants (Gagliano et al., 2017), but there is no evidence of how they react to the sounds of drones.

However, the SPADE project does not take a deep level of user engagement and does not take a participatory design approach, or at least not in each pilot. This decision signals to the evaluators that the proposed solutions might discriminate against some of the agents. In the impact research, we will look back and study the lists of participants and methods used in the design phase and evaluate how inclusive they were.

Work simultaneously with (new) ethical guidelines

By applying "gender innovation" goggles, the ethical guidelines and this framework will serve as a tool to monitor ethical issues related to humans, personal data, animals, environment, health and safety, and artificial intelligence during the research project. We created a logbook on the basis of this framework/matrix and kept logs/notes during all the work meetings to collect new hypotheses, questions, and actions in the design phases that were undertaken to avoid discriminatory biases. In the early stages, when we drafted the guidelines, we learned even more about certain concerns, enhancing our discussions on gender dimension issues.

Rethink research questions, reference models, and language, and reflect on values

In general, it is known that research questions and language in, for example, interacting directly with users (through interviews and surveys) may be biased because of gender dynamics. Even in SPADE, there are discussions about what should be the outcomes and impacts of a digital solution. For digital agriculture, for example, developers may have quite narrow values regarding what good farming is, which has been shown to lead to design choices and data selection choices that privilege large-scale farmers over smaller ones or focus more on quantity of a uniform product (such as tomatoes) (Van Der Burg et al., 2021). There is also interviewer bias: Gender and age bias can influence diagnostic questions in health, but also in cocreation; women often answer what they think other women would answer to be perceived as "feminine," even if they really like STEM.

As SPADE involves a lot of information, different data formats, needs, and requirements, the partners work with reference models and trustworthiness frameworks. Researchers at this stage are investigating reference models and standards, looking at Specific Operations Risk Assessment (SORA) for drones. Regarding trustworthiness, in the research on, for example, drones, women were reported to be more worried about malfunction and intentional misuse, privacy, and other risks (Sabino et al., 2022). The trustworthiness framework may discourage women from participating if it does not incorporate women's subjective logic and concerns as well. It is therefore imperative to record the concerns of women and all diverse groups of people and to incorporate them into frameworks.

Lastly, the DT is "a dynamic virtual representation of a physical object or system, usually across multiple stages of its lifecycle, that uses real-world data, simulation, or machine learning models combined with data analysis to enable understanding, learning, and reasoning. DT can be used to answer what-if questions and should be able to present insights in an intuitive way" (Clark et al., 2019). As they are a representation, this might encompass many biases and worldviews of the people designing the DT.

In agriculture, DTs are at a low maturity level, and often what gets twinned are not living objects (like livestock), but drones, tractors, and the landscape (Pylianidis et al., 2021).

Lastly, it is wise to avoid reinforcing gender stereotypes in the design. In many designs, gender stereotypes are (unconsciously) integrated. It is not advised to create a separate website/interface or drone that is pink to make it more "woman-friendly" or "women-attractive," first because it reinforces the idea that women like pink, secondly because often these women-attractive markets/products are more expensive although they sell the same (e.g., shaving materials), and thirdly because it does not address deeper and structural problems. The SPADE project will not be able to solve structural challenges (see Contribute to reducing structural changes), but can avoid fostering gender stereotypes at all design stages.

Reflect on data access, and the influence on and of power relations and distribution of benefits, costs, and risks

With digital innovation, like drones, we are improving access and regulating governance to data and information to make better decisions that can increase economic benefits, not just for a certain group (e.g., rich, digitally savvy men), as well as decrease risk at the same time as reducing costs. In the later stages of a project like SPADE, different partners will look at implementation in the real world (through pilots), exploitation, and the market uptake of the developed solutions. However, there are ethical concerns when using DTs because they have diverse socioeconomic effects (Van Der Burg et al., 2021). Access to technology may thus result in a competitive advantage (or the disappearance of it) for certain users. Based on previous historical analyses, mostly in countries in Africa and Asia, women often have less access to microcredits (Mwangi and Kariuki, 2015; Gurung and Bisht, 2014). In addition, the implementation of DTs may shift power relations among actors in the agri-food chain, depending on who uses the DTs and who has access to the data and knowledge produced with DTs (Van Der Burg et al., 2021), so there might also be actors sabotaging the implementation if they see it threatening their power or independence.

For wider social adoption across gender

Especially men were more likely to accept drones; however, that was related to their knowledge about drones as most drone operators are men (Aydin, 2019). Most of these activities are related to disseminating, exploiting, and communicating in order to have a wider social impact, which means a wider social adoption. Social networks are an important factor in social acceptance. One of the factors SPADE will collect in its surveys of potential users is the presence of social networks and access to the internet. Resources, such as information about innovations, are not only business driven but also facilitated via intermediary governance models.

Work with women-only spaces and events and promote women panels

From a governance perspective, it is necessary to understand which factors can lead to a higher likelihood of adoption of upcoming digital technology, like a DT or drone. For example, based on research on the adoption of drones in agriculture in China (Wachenheim et al., 2021), males were more likely to adapt, especially if they participated in community organizations. Age or education level was not significant. A major reason is that farmers and foresters lack awareness and insight and rely on consultants from intermediary organizations and networks for assistance. Additionally, these community organizations can also raise capital to buy this expensive technology and make it possible for farmers and foresters to use it. One could argue that these networks are not only helpful for knowledge distribution but can also be the pillars of enabling a reasonable benefit-sharing, risk-sharing, and profit-distribution model. Like we observe with other expensive machineries and technologies, like tractors, etc., we need to explore which settings enable a sharing business or cooperative model for drones. However, in certain contexts, especially in still conservative spaces and places, where men and women move in separate networks (as in Japan and China), this can lead to exclusion. Women do not have access to information about new technologies. We see that women often create their own networks and groups to help each other (e.g., Kiptot & Franzel, 2012; Laszlo Ambjörnsson, 2021). Hence, gender-sensitive innovation projects should identify and locate them too when they look for end users to test the quality of a new product/service/innovation. If they do not exist, this should be signaled to policy makers and other instances to create (more) spaces and events for female farmers and foresters.

In the SPADE project, when we collect opinions and concerns about drones and DTs, we plan to have a diversity of intermediate groups and domain experts. The work package related to dissemination, exploitation and communication (DEC) activities will make extra efforts to locate female forestry and agriculture associations in Europe. This might involve or lead to the creation of women-only spaces. Some SPADE partners have already experienced and seen the effect of women-only events for developers, for example, as they sense that women are more comfortable in a space where they do not feel like a minority.

As long as gender-based violence and discrimination exist, women-only spaces are needed to ensure a smooth transition. Critics claim they do not address the issue radically and keep men separate from women's needs. In some countries, as in Norway, these spaces are often postfeminist and do not see the need for "gender equality actions" (Laszlo Ambjörnsson, 2021). It also should be noted that there are remarks and critiques about the term gender mainstreaming, which may be transformative in theory, but in practice, "it has tended to make women the subject of change, where the goal is to fit women into the status quo rather than transform the status quo" (Shortall & Bock, 2015, p. 663 in Laszlo Ambjörnsson, 2021).

Create nearby and virtual spaces/events and record them

One way to transform the status quo is to accommodate the needs and schedules of diverse people. In dissemination and communication activities that contribute to social adoption, we will also take measures that are inclusive of diverse everyday schedules. Many parents and farmers who have animals are bound by schedules and the tantrums and wishes of children and animals.

Organize virtual workshops and make them available online — so more people can see and learn about them at the time that fits them. This will benefit mothers (and fathers) who have less flexible schedules. In a rural setting, it is mostly women who are bound to these schedules (e.g., feeding the kids).

A posteriori — impact assessment

Another part of the horizontal gender component in the SPADE project consists of interviews or surveys with individual foresters, farmers, and representatives of large forest companies and the aforementioned intermediary groups and domain experts and — if possible to locate them — people who left the forestry before retirement to learn how gender norms advance or impede the adoption of the proposed drone technology in agriculture and natural resource management across different contexts and social structures. Questions that should be asked — among others — include the following:

How and where do these drone technologies advance or weaken women's agency and change gender norms to the benefit or harm of women?

When and under which circumstances do changes in gender norms and women's and men's agencies catalyze innovation and promote inclusive outcomes?

Are there positive norms that we should recognize and strengthen? What contextual factors influence this relationship?

Based on newly available evidence a posteriori, should we rethink the standards and reference models that we used in the first years of SPADE? (inspired by Petesch et al., 2017).

Participants will be asked about their biological sex and their gender identity. This implies clear operational rules for evaluators regarding who is classified as a woman, man, or gender-fluid or nonbinary person. It is also recommended to document how many female and male participants were present and spoke, and describe the dynamics (who spoke and how much of the time) (Søraa et al., 2020).

Another impact research engaged with social constructivist theories (e.g., the agent network by Bruno Latour), which acknowledges that drones, sensors, DTs, goats, sheep, and crops are also changing the behavior and preferences of humans and each other. Therefore, an impact analysis should include the study of structures, like policies, regulatory frameworks, and the relationships between the different agents involved in the SPADE project. This can help us to understand how users (farmers and foresters) adapt to the proposed solutions, how policy makers situate innovation within existing legal, i.e., structural, frameworks, and, of course, the limitations that both technology and nature have on the actual development possibilities seen from the developers' side. In the end we have to evaluate how the developed solutions (will) influence gender and other social norms on access to and benefits from resources such as (micro)credits, land ownership, training, mobility, decision-making power, and support from family and community members (Cole et al., 2015; Badstue et al., 2018).

Contribute to reducing structural changes

Some structural challenges cannot be reduced by a project like SPADE. For example, SPADE engages with domains and primary sectors where there is a strong gender divide, even in more egalitarian societies as in Norway (e.g., Seddighi (2022)). Although Norway has one of the highest gender equality indexes in the world, women are still underrepresented in forestry. In Norway, half of the forests are owned by large companies and the other half by individuals, of which only 25% are women, while the proportion of female board members and employees in the larger forest owners' associations is between 25% and 50% in companies (Nordic Forest Research 2022). Even entry points into agriculture and farming via ICT work are difficult for women (and also for men) if they have moved away to study in cities. However, women succeed in ICT jobs in rural areas as the ICT sector is growing, but the thinness of organizations can have both positive and negative sides, and this negative side (small network) can lead to women moving out again (Seddighi, 2022). Working in these sectors and areas comes with a number of structural challenges.

Lack of representation, recognition, and participation and uneven distribution of risks, costs, and benefits are systemic and cannot be solved by fragmented R&I projects. However, there should be a responsibility to communicate findings, and especially patterns, to policy makers, planners, and other actors in influential positions. SPADE will communicate about the demonstration effects (e.g., effects as a result of pilot projects) − including positive and negative effects − especially negative, because this is also critical information for EU policy makers.

It is encouraging to see the European Commission pushing for more analysis of the gender dimension and the involvement of social scientists in research, development, and research projects under their programs like Horizon Europe. In the coming years, these gender dimension studies will contribute to more gender- and diversity-sensitive policies and projects. The next step is to set a minimum budget quote for gender and justice research tasks in all projects, so they do not remain small tasks performed by one or a few researchers. We did not find a lot of research and had to be creative in finding enough data sources to write this chapter, which also provides the background for the gender dimension task in the SPADE project. A signal that we need more intense empirical data and research.

By fostering a critical examination of gender roles, norms, and relationships, we should recognize and strengthen positive norms that support equality. We have to transform the underlying social structures, policies, and widely held beliefs that perpetuate gender inequality. Instead of supporting method-driven projects, like the further development of drones and data governance, budget should be allocated to projects in the digital humanities to create (ongoing) campaigns that strengthen positive norms and debunk negative norms (Table 11.1).

Summary

The following table is a summary of a series of virtual roundtables. They echo the recommendations elaborated upon in Section 3. Ideally, we will return to this exploratory research and reflections at the end of or after the SPADE project (2026) in order to determine what worked and whether this framework and list of recommendations were useful — to what extent and to whom.

Table 11.1 Summary of the round tables on the gender perspective at SPADE's beginning about concerns and possible measures.

Solutions	Concerns — design and impact	Responses and possible mitigation measures
Impacts on rural areas	Women will not reap the benefits, because of underrepresentation and structural societal expectations that discourage them from doing ICT-related trainings. Tension between feminist and postfeminist ideas that prefer women to act like men instead of changing society in order to care for diverse needs and schedules, and critique gender mainstreaming concepts. Tunnel vision on gender (and/or biological sex), and not on research of how innovation will influence gender and other social norms on access to and benefits from resources such as (micro) credits, land ownership, training, mobility, decision-making power, and support from family and community members (Bock and Shortall, 2017).	Do proactive participatory events and surveys in deployment areas to collect the needs of women in rural areas (developers, farmers, and foresters). Make user-friendly interfaces. Collect gender-segregated data about the usage and benefits that users experience from deploying the solution. Investigations of gender performances in forestry and agriculture. Intersectionality analysis, and collect data about age, access to networks, etc., and analyze combinations of these segregated data sets.

(*Continued*)

Table 11.1 (Continued)

Solutions	Concerns – design and impact	Responses and possible mitigation measures
Drones and digital twins	Female and male animals have different behaviors and are likely to react differently to sensors and/or drones. The visual recognition experts reflect on the gender dimension.	Likely, there will be no DT of animals (there are limited projects on living objects in agriculture), and therefore no distinction between female and male animals. Developing visual recognition algorithms to distinguish ewes from rams.
Communication and education activities (WP5)	DEC activities do not reach a diverse group of (potential) users. Signal structural challenges that a project like SPADE cannot address to policy makers and bodies (including standardization bodies, etc.).	Make lists of women networks and intermediaries (owner associations), especially in the deployment areas of the pilots. Experiment with women-only events. Have enough women on panels.
Project management (WP6)	Uncertainty about how to document the gender dimension of SPADE in an innovative and refreshing way. Train people in inclusive facilitation and management.	Logbook creation reporting the published papers or reports, including participants asked about gender and checked if that was a relevant variable or not.

Acknowledgments

This work has been supported by the SPADE project, funded by the European Union's Horizon Europe Research and Innovation programme within HORIZON-CL6-2021-GOVERNANCE-01 under Grant Agreement no. 101060778.

We would also like to acknowledge the SPADE partners, especially Pierrick Le Guillou (AnySolutions), Aristotelis Tagarakis (CERTH), Costas Davarakis (NST), Christian Weigel (Fraunhofer), Fleming Sveen, Asbjørn (Hafenstrom), Raquel Almeida and Mariia Levanchuk (SPI), and Philippe Krief (OpenSource) for sharing data, opinions, and their experiences in identifying and mitigating concerns.

References

Aydin, B., 2019. Public acceptance of drones: knowledge, attitudes, and pEractice. Technol. Soc. 59, 101180.

Badstue, L.B., Petesch, P., Feldman, S., Prain, G., Elias, M., Kantor, P., 2018. Qualitative, comparative, and collaborative research at large scale: an introduction to GENNOVATE. J. Gend. Agric. Food Secur.

Bock, B., Shortall, S., 2017. Gender and rural globalisation: an introduction to international perspectives on gender and rural development. Gender and Rural Globalisation: International Perspectives on Gender and Rural Development.

Butler, J., 1990. Subversive bodily acts' in gender trouble. Feminism Subversion Identity 79−141.

Clark, A.S., Schultz, E.F., Harris, M., 2019. What are digital twins. Technical Report. IBM. https://developer.ibm.com/articles/what-are-digital-twins.

Cole, S.M., Puskur, R., Rajaratnam, S., Zulu, F., 2015. Exploring the intricate relationship between poverty, gender inequality and rural masculinity: a case study from an aquatic agricultural system in Zambia. Culture, society & masculinities 154−170. Available from: https://doi.org/10.3149/CSM.0702.154.

Crenshaw, K., 1989. Demarginalizing the intersection of race and sex: A black feminist critique of antidiscrimination doctrine, feminist theory and antiracist politics. u. Chi. Legal f., p. 139.

Elias, M., Mudege, N.N., Lopez, D.E., Najjar, D., Kandiwa, V., Luis, J.S., et al., 2018. Gendered aspirations and occupations among rural youth, in agriculture and beyond: a cross-regional perspective. J. Gend. Agric. Food Secur.

European Commission, 2022. SPADE's Factsheet - Multi-purpoSe Physical-cyber Agri-forest Drones Ecosystem for Governance and Environmental Observation. Available from: https://doi.org/10.3030/101060778.

European Union, 2020. Gendered Innovations 2: How Inclusive Analysis Contributes to Research and Innovation. Available from: https://research-and-innovation.ec.europa.eu/news/all-research-and-innovation-news/gendered-innovations-2-2020-11-24_en, https://doi.org/10.2777/316197Accessed 22 December 2022.

Gagliano, M., Ryan, J.C., Vieira, P. (Eds.), 2017. The Language of Plants: Science, Philosophy, Literature. U of Minnesota Press.

Gurung, D.D., Bisht, S., 2014. Women's empowerment at the frontline of adaptation: emerging issues, adaptive practices, and priorities in Nepal. In: ICIMOD Working Paper 2014/3. Kathmandu, Nepal.

ISO, 2022. Women in Standards. https://www.iso.org/strategy2030/key-areas-of-work/diversity-and-inclusion/women-instandards.html. Accessed 22 December 2022.

Jenkins, K., McCauley, D., Heffron, R., Stephan, H., Rehner, R., 2016. Energy justice: a conceptual review. Energy Res. Soc. Sci. 11, 174−182.

Kiptot, E., Franzel, S., 2012. Gender and agroforestry in Africa: a review of women's participation. Agrofor. Syst. 84 (1), 35−58.

Klauser, 2018. Surveillance farm: towards a research agenda on Big Data agriculture. Surveill. Soc. 16 (3), 370−378.

Krishna, K.R., 2017. Push Button Agriculture: Robotics, Drones, Satellite-Guided Soil and Crop Management. CRC Press.

Laszlo Ambjörnsson, E., 2021. *Gendered Performances in Swedish Forestry: Negotiating Subjectivities in Women-Only Networks* (Doctoral dissertation). Department of Human Geography, Stockholm University.

Macinnis-Ng, C., Zhao, X., 2022. Addressing gender inequities in forest science and research. Forests 13 (3), 400.

Manfre, C., Rubin, D., Allen, A., Summerfield, G., Colverson, K., Akeredolu, M., 2013. Reducing the gender gap in agricultural extension and advisory services: how to find the best fit for men and women farmers. Meas. Brief. 2, 1−10.

Mosse, D., 1994. Authority, gender and knowledge: theoretical reflections on the practice of participatory rural appraisal. Dev. Change 25 (3), 497−526.

Mwangi, M., Kariuki, S., 2015. Factors determining adoption of new agricultural technology by smallholder farmers in developing countries. J. Econ. Sustain. Dev. 6 (5).

Nikulina, V., Lindal, J.L., Baumann, H., Simon, D., Ny, H., 2019. Lost in translation: a framework for analysing complexity of co-production settings in relation to epistemic communities, linguistic diversities and culture. Futures 113, 102442.

Nordlund, A., Westin, K., 2011. Forest values and forest management attitudes among forest owners in Sweden. Forests 2 (1), 30−50. Available from: https://doi.org/10.3390/f2010030.

Ortner, S.B., 1972. Is female to male as nature is to culture? Fem. Stud. 1 (2), 5−31.

Oudshoorn, N., Neven, L., Stienstra, M., 2016. How diversity gets lost: age and gender in design practices of information and communication technologies. J. Women Aging 28 (2), 170−185.

Petesch, P., Badstue, L.B., Prain, G., Elias, M., Tegbaru, A., 2017. Entry Points for Enabling Gender Equality in Agricultural and Environmental Innovation. GENNOVATE resources for scientists and research teams. CDMX, Mexico.

Petesch, P., Badstue, L.B., Prain, G., 2018. Gender Norms, Agency, and Innovation in Agriculture and Natural Resource Management: The GENNOVATE Methodology.

Plumwood, V., 1993. Feminism and the Mastery of Nature. Routledge, London and New York.

Pylianidis, C., Osinga, S., Athanasiadis, I.N., 2021. Introducing digital twins to agriculture. Comput. Electron. Agric. 184, 105942.

Sabino, H., Almeida, R.V., de Moraes, L.B., da Silva, W.P., Guerra, R., Malcher, C., et al., 2022. A systematic literature review on the main factors for public acceptance of drones. Technol. Soc. 102097.

SCC, 2020. When One Size Does Not Protect All: Understanding Why Gender Matters for Standardization - https://www.scc.ca/en/about-scc/publications/general/when-one-sizedoes-not-protect-all. Accessed at 22 December 2022.

Schiebinger, L., 2021. Gendered innovations: integrating sex, gender, and intersectional analysis into science, health & medicine, engineering, and environment. Tapuya: Latin Am. Sci. Technol. Soc. 4 (1), 1867420. Available from: https://doi.org/10.1080/25729861.2020.1867420.

Schiebinger, L., Klinge, I., Sánchez de Madariaga, I., Paik H.Y., Schraudner, M., & Stefanick, M.L. (Eds.) 2011-2021. Gendered Innovations in Science, Health & Medicine, Engineering and Environment. http://genderedinnovations.stanford.edu/index.html.

Seddighi, 2022. Women's careers in ICT in Norwegian rural contexts Women in Technology Driven Careers. In: Conference Paper. - https://www.vestforsk.no/nn/publication/womens-careers-ict-norwegian-rural-contexts. Accessed at 19 April 2023.

Shortall, S., Bock, B., 2015. Introduction: rural women in Europe: the impacts of place and culture on gender mainstreaming the European Rural Development Programme. Gend. Place Cult. 22 (5), 662−669. Available from: https://doi.org/10.1080/0966369X.2014.917819.

Snyder, H., 2019. Literature review as a research methodology: an overview and guidelines. J. Bus. Res. 104, 333−339.

SPADE Website, 2023. https://spade-horizon.eu/.

Søraa, R.A., Anfinsen, M., Foulds, C., Korsnes, M., Lagesen, V., Robison, R., et al., 2020. Diversifying diversity: inclusive engagement, intersectionality, and gender identity in a European Social Sciences and Humanities Energy research project. Energy Res. Soc. Sci. 62, 101380.

Søraa, R.A., Vik, J., 2021. Boundaryless boundary-objects: digital fencing of the CyborGoat in rural Norway. J. Rural. Stud. 87, 23–31.

Tannenbaum, C., Ellis, R.P., Eyssel, F., Zou, J., Schiebinger, L., 2019. Sex and gender analysis improves science and engineering. Nature 575, 137–146.

UN Women, 2014. World Survey on the Role of Women in Development. UN Women, New York, p.40.

Umaerus, P., Högvall Nordin, M., Lidestav, G., 2019. Do female forest owners think and act "greener"? For. Policy Econ. 99, 52–58. Available from: https://doi.org/10.1016/j.forpol.2017.12.001.

van Hoof, J., 2015. Female thermal demand. Nat. Clim. Change 5 (12), 1029–1030.

Van Der Burg, S., Kloppenburg, S., Kok, E.J., Van Der Voort, M., 2021. Digital twins in agri-food: societal and ethical themes and questions for further research. NJAS: Impact Agric. Life Sci. 93 (1), 98–125. Available from: https://www.tandfonline.com/doi/pdf/10.1080/27685241.2021.1989269.

Vate-U-Lan, P., Quigley, D., Masouras, P., 2016. Smart Dairy Farming Through the Internet of Things (IoT).

Wachenheim, C., Fan, L., Zheng, S., 2021. Adoption of unmanned aerial vehicles for pesticide application: role of social network, resource endowment, and perceptions. Technol. Soc. 64, 101470.

Wajcman, J., 2010. Feminist theories of technology. Camb. J. Econ. 34 (1), 143–152.

Weber, R.N., 1997. Manufacturing gender in commercial and military cockpit design. Sci. Technol. Hum. Values 22 (2), 235–253.

WHO (World Health Organization), 2002. Gender Analysis in Health: A Review of Selected Tools. WHO, Geneva.

Yam-Viramontes, B., Cardona-Reyes, H., González-Trejo, J., et al., 2022. Commanding a drone through body poses, improving the user experience. J. Multimodal User Interfaces 16, 357–369. Available from: https://doi.org/10.1007/s12193-022-00396-0.

CHAPTER 12

An overview of unmanned aircraft systems (UAS) governance and regulatory frameworks in the European Union (EU)

Anssi Rauhala, Anne Tuomela and Pekka Leviäkangas
Civil Engineering, Faculty of Technology, University of Oulu, Oulu, Finland

Introduction

The early use of unmanned aircraft (or aerial) systems (UAS) was mostly limited to the military and defense sector. However, the use of UAS has become also widespread in commercial and recreational applications. According to Scopus searches by Nex et al. (2022), more than 80,000 papers with "drone," "UAS," or similar synonym in the title or keywords have been published since 2001. Market forecasts also indicate promising developments on civil UAS applications. The military and defense subsegments dominate the market, but multiple forecasters predict that the agriculture subsegment will have the fastest growth in the 2020s (e.g., Fortune Business Insights, 2020; Precedence Research, 2022; Researchdive, 2022). For example, Fortune Business Insights (2020) forecasts that in the 2019−27 period, agriculture subsegment will have the strongest growth with a compound annual growth rate (CAGR) of 18.1%, rising from 1.02 billion USD to 3.7 billion USD.

Although the advancements and availability of civil UAS has the potential to bring considerable benefits in fields such as surveying, inspections, search and rescue, security, and of course, agriculture, it can also pose threats and risks to, for example, physical safety, privacy, and data security. To tackle the risks associated with civil UAS and concerns about privacy and public safety, first national regulations begun to be promulgated in European Union (EU) in 2002−03 (Stöcker et al., 2017). On EU level, the early Regulation (EC) 2008/216 only applied to UAS with an operation mass of over 150 kg,

Unmanned Aerial Systems in Agriculture
DOI: https://doi.org/10.1016/B978-0-323-91940-1.00012-8

which left much of the regulatory powers regarding smaller UAS to the hands of the EU member states. Consequently, EU member states largely continued to set their own policy with regards the use of UAS (Tsiamis et al., 2019), which led to fragmented and heterogenous set of national level rules with varying levels of implementation (Stöcker et al., 2017).

The European Parliament resolution on safe use of remotely piloted aircraft systems (2014/2243(INI)) noted that the lack of harmonized rules at EU level might impede the development of a European drone market, and stressed the needs to protect privacy, safety, and security. The resolution further supported the development of a clear and harmonized risk-based regulatory framework, and development of technology and standards to enable integration of UAS into European airspace, while also noting that public acceptance is a key to growth of UAS services. At the end of long process and public consultations, the UAS sector strategy and regulatory framework in EU was radically changed by replacing the fragmented and dual approach of Reg. (EC) 2008/216 by the centralized, top-down framework of Regulation (EU) 2018/1139 on common rules in the field of civil aviation (Pagallo and Bassi, 2020). This new regulation further mandated the Commission to adopt delegated and implementing acts to regulate the sector.

The European Aviation Safety Agency (EASA) also has a mandate to develop technical rules upon the request of the European Commission and soft powers to, for example, issue Guidance Materials and Acceptable Means of Compliance on how to comply with the new regulations (Pagallo and Bassi, 2020). EASA can also develop standards on how to integrate UASs to the European single airspace strategy and coordinate actions of member states with regards to, for example, certifications. The soft powers to, for example, develop standards has been described as a middle ground between top-down regulatory approach and the forces of the market (Pagallo and Bassi, 2020).

What is governance?

"Governance is the exercise of political, economic and administrative authority necessary to manage a nation's affairs" (OECD). *"The process by which decisions are made and implemented (or not implemented). Within government, governance is the process by which public institutions conduct public affairs and manage public resources"* (IMF). Governance defines the way things are managed, and who is mandated or obligated to manage them. Therefore governance is a "rule book" for decision-making and maintaining responsibilities and accountability.

While market forecasts indicate promising growth for UAS business and services, this can also be largely dependent on agile governance to harness innovation. While regulators typically resort to traditional top-down approach, developers, operators, and manufacturers often call for flexible bottom-up approach to enable rapid development (Du and Heldeweg, 2019; Hernández and Amaral, 2022). On the other hand, constantly evolving and changing regulatory landscape also nibbles away the predictability of governance, which is often seen as one requirement for investment in a sector. Furthermore, it certainly makes entry in the field more demanding when even existing operators have trouble keeping up to date with the development. For example, the State of the Drone Industry Report 2022 by DroneDeploy (2022) highlights that 42% of respondents (766 customers from 40 + countries) in 2022 report having trouble keeping track of the changing laws and regulations, compared to only 25% respondents in 2020. The typical regulatory approach to predominantly focus on controlling safety risks also runs the risk of overlooking broader societal impacts of UAS (Hernández and Amaral, 2022). In the end, the flourishing development of the UAS industry can be also largely dependent on the social acceptability of wider use of UAS in various applications.

ICAO Model UAS Regulations.

ICAO (International Civil Aviation Organization) has provided, after the request of its members, a regulatory framework for UAS. These regulations are titled as Parts 101, 102, and 149. In essence, Part 101 states that all unmanned aircrafts (UA) should be registered, but UAS weighing less than 25 kg requires no operational review. However, the regulation also states that if the drone weighs more than 15 kg, it must be inspected and approved according to certain procedures stated in Part 101 and Part 102. Part 102 covers UAS that weigh more than 25 kg and requires a certificate for operation. Also 102 allows a quicker review of the type of UA when there are credible declarations of the manufacturer or if the UA has been approved by an approved aviation organization (AAO), for example, by a national aviation regulating body or by an organization that has been delegated to these types of tasks. Part 149 is about the role of AAOs to serve as designee authorized by the CAA. Basically, the global UA regulatory frameworks dictate the regional and national regulations but preparing the regulations or observing the needs for new regulations does not always follow top-down process. For example, EASA has in many cases been the driver also for global regulations, having identified the European statutory needs.

Overview of current European regulations

As mentioned, UAS operations in EU currently fall under the so-called "Basic Regulation" (Regulation [EU] 2018/1139) on common rules in the field of civil aviation. Unlike the previous Reg. (EC) 2008/216, the new Reg. (EU) 2018/1139 applies to all civil UAS regardless of size and weight with only few exceptions, such as tethered aircraft with a maximum take-off mass of no more than 1 kg. The new regulation also reduces the powers of member states and devolves most of the ruling powers to European Commission and EASA. Some exceptions are still admitted to the centralized model, such as possibility for member states to define, for example, permissive, and restrictive zones for UAS operations in their territory, and ability for member states to grant time-limited exceptions from nonessential requirements in the event of unforeseeable circumstances (Bassi, 2020).

The Implementing Reg. (EU) 2019/947 lays down detailed provisions for the operation of UAS as well as for personnel and organizations, whereas the Delegated Reg. (EU) 2019/945, for example, lays down the requirements for the design and types of UAS intended to be operated under the rules of Reg. (EU) 2019/947. Delegated Reg. (EU) 2019/945 also lays down rules for third-country operators, when conducting operations pursuant to Reg. (EU) 2019/947 within the European single sky airspace. Notably, the technical and legal requirements set out in the regulations follow the distinction suggested by EASA Opinion 01/2018 (Bassi, 2020), which categorizes civil UAS operations into "open," "specific," and "certified" categories, and sets boundaries to which type of UAS can be operated in which types of operational contexts.

Open category specifies low-risk UAS use cases and does not require special permits or licenses, and there is no need for operational authorization of any kind. However, most UASs operable in the open category require registration as a UAS operator and passing an online training and examination defined by the competent authority (see Table 12.1). The open category has the following general requirements that must be met:
- maximum attainable height above ground or sea level must not exceed 120 m,
- the UAS must have max. take-off mass of less than 25 kg,
- the operation must be carried out on visual line of sight (VLOS),
- no carrying of dangerous goods or dropping of items during the flight,

Table 12.1 Subcategories A1—A3 for the open category unmanned aircraft systems (UAS) operations after the transition period (January 1, 2024 onwards).

A1	A2	A3
UAS mass <900 g CE markings: C0 and C1	UAS mass <4 kg CE markings: C2	UAS mass <25 kg CE markings: C2, C3, C4
Flights are permitted in densely populated areas above individual people, but not over assemblies of people	Flights are permitted in densely populated areas at safe distances (30 or 5 m with an active low speed mode function) from uninvolved people	Flights are permitted in sparsely populated areas and far away from people and settlements (150 m from residential, commercial, industrial, or recreational areas)
UAS mass <250 g: no training/examination requirement. UAS mass 500—900 g: online theoretical examination	Online theoretical examination + additional theoretical knowledge examination	Online theoretical examination

- UAS must keep safe distance to people and must not be flown above assemblies of people,
- the operation must take danger areas of aviation and UAS geographical zones into account.

UAS operations in the open category are further divided into subcategories A1—A3 that give more limitations depending on the UAS take-off mass (Table 12.1). Starting from January 1, 2024, UASs operated in the open category must also bear a class identification marking (C0—C4) or be privately built or be purchased before December 31, 2023.

The competent authority can also designate separate UAS geographical zones in which UAS operations are exempt from one or more of the requirements of open category, such as the maximum take-off mass of 25 kg or the requirement of the pilot to operate only one UAS at a time. If the operation cannot be performed in accordance with the open category terms, separate specific category authorization based on a risk assessment is required from the national competent authorities. There are four available options to operate in the specific category: a declaration based on a national or European standard scenario (STS), obtaining an authorization based on a predefined risk assessment (PDRA) or a specific operations risk assessment (SORA), or having a light UAS operator certificate (LUC).

There are currently two European level STSs defined: STS-01 for a VLOS operation over a controlled ground area in a populated environment with a UAS bearing C5 class identification, and STS-02 for a beyond visual line of sight (BVLOS) operation with airspace observers over a controlled ground area with a UAS bearing C6 class identification. The launch date for European STSs will by the current knowledge be January 1, 2024 since drones bearing class identifications C5 or C6 required for the scenarios are not currently available. The STSs do not require authorization as they are based on declaration that the operator fulfills the requirements of the scenario, and, for example, has the required certificate for theoretical knowledge, accreditation of completion of STS practical skills training, and also has an operation manual that is required for all the specific category operations.

In the PDRA-based approach, the authorities have already carried out the risk assessment for the most common types of operations, thus making the process more streamlined compared to SORA process as the operator can simply prove that they comply with the given boundary conditions set in the PDRA and submit an application to the national aviation authority (NAA). If the operation is not covered by any STS or PDRA scenarios, the operator must conduct a SORA and identify possible risk mitigations to comply with safety objectives. The SORA is a 10-step approach following the guidelines provided by JARUS (2019).

The LUC focuses on organizational approval instead of individual operation, thus granting the organization ability to self-authorize STS operations without a declaration and/or PDRA and/or SORA operations without an application. In addition to the operations manual required by all specific category operations, the acquisition of a LUC further requires the operator to have, for example, a safety management system, a documentation system, and a LUC safety manual. In an interview by Öz et al. (2022), representative of the Finnish competent authority states that even applying for an operational authorization is challenging for many operators and granting a LUC is not currently realistic action in many cases as multiple successful authorizations are typically needed to demonstrate organizational maturity.

In the certified category, the risks of UAS operating are considered high and comparable to manned aviation, and the UAS can be operated only by certified and licensed operators. In most agriculture-related use cases, the open and specific categories are most relevant and interesting since these use cases allow relatively flexible adoption of UAS. Interview of a European level aviation safety authority by Öz et al. (2022) indicates that in the long-term STSs and PDRAs along with the open category should cover most operations, whereas SORA-based authorization will be an exception. However, the adoption of EASA regulations is still in its early phases, so the current categorization may not be permanent in the medium or long run.

There are already notable differences between some EU countries how the competent authorities define the operational environment requirements, registering of operators, training, and applications to be authorized as operators. Table 12.2 highlights these differences (as of January 2023). The differences relate to format of operating airspace visualization, languages available, information clarity, etc. Based on quick review, none of the national sites were too easy to navigate and the sites differed substantially in terms of user interfaces. The national EASA rule implementation and application differences may well fulfill the letter of the requirements, but from the user experience perspective the quality could be assessed as "satisfactory," at best.

International governance architecture for aviation in a nutshell.

Aviation is governed at international level because much of the aviation activities are carried out across national borders, from one country to another. Both people and goods are transported in increasing volumes. The governance starts from United Nations specialized agency, ICAO, that comprises almost 200 member countries (193 member states for 2022—25 governance period). The mission is to *"develop policies and Standards, undertakes compliance audits, performs studies and analyses, provides assistance and builds aviation capacity through many other activities and the cooperation of its Member States and stakeholders."* The tasks of ICAO are multiple, extending from research and innovation to technical advising, keeping global statistics, and building scenarios and forecasts. One of the important areas of activity is aviation safety. As an international standard setting and governance body, ICAO defines many important processes, rules, and recommendations that bind its members, once these statutory elements have been approved by ICAO decision-making bodies. The regional governance bodies comprise, for example, United States Federal Aviation Administration (FAA) and EASA that have legislative powers. In Asia, such a regional organization does not exist. At least ICAO and EASA have regional offices in Asia. Both EASA and FAA are considered as Civil Aviation Authorities (CAA) of mega-scale, meaning that they

(Continued)

(Continued)

have a wider mandate than that of one country or state. The European aviation governance system includes basically EASA, EUROCONTROL, and the
national aviation regulators and authorities. While EASA focuses on aviation
safety, EUROCONTROL has been established to support the European Single
Market for aviation, both civil and military aspects of aviation. EASA is essentially a regulating body, whereas EUROCONTROL adopts more of a coordina-
tor role in European aviation and does not possess regulatory powers. The
division of mandates is not always clear, and one signal is that a new joint
body, EASA-EUROCONTROL Technical and Coordination Office (TeCo) was
established in July 2020. EASA has the EU member states' national aviation
authorities and regulators as their country representative members.
EUROCONTROL has a wider membership composition, including member
countries outside the EU, for example, from North Africa and Eastern Europe.
IATA, the International Air Transport Association, is a global industry organization, tasked to work for the benefit of airline companies. Its role is obviously
relevant when international regulations are prepared. The global governance
architecture for aviation is illustrated in Fig. 12.1.

Challenges and social acceptability

Despite the already rather complex set of regulations, several issues remain
open in fields such as data protection and privacy, tortuous liability, criminal justice, telecommunication, and cybersecurity law, albeit many of
these issues are not up to power of EASA or competent authorities (Bassi,
2020). Then again, some of these fields, such as tortuous liability, fall
within regulatory powers of member states, thus raising the risk of frag-
mentation (Pagallo and Bassi, 2020). Further issue is what kind of balance
should exist between the risk-based approach of UAS regulations and the
data protection impact assessment (DPIA) of the General Data Protection
Regulation (GDPR), Reg. (EU) 2016/679, which requires data controllers to assess risks and mitigating measures for the protection of personal
data (Bassi, 2020).

The risk-based approach of UAS regulations requires real world
experiences and data for its development. With regarding the U-space
concept, EASA has noted the lack of data for thorough safety risk assess-
ments and the lack of common data exchange infrastructure. There is a
clear need for sufficient UAS flight safety incident and occurrence reporting and data exchange system, which can support identifying problems in

Table 12.2 Examples of national adoption of European Aviation Safety Agency (EASA) unmanned aircraft systems regulations.

	Austria	Bulgaria	Croatia	Finland	Greece
Available airspaces	Relatively clear overall map with restricted air zones; detailed map after defining the location	Maps available in Bulgarian only, other information also in English; the maps can be zoomed with enough detail	Maps available with different zone restrictions	Maps shown where drones are permanently forbidden; a list of special zones given that need to be clicked separately	No links, nor information through EASA site
Registering	Registration via forms to be emailed to Austro Control (NAA)	Registration platform only has latest news section in the English version; the Bulgarian version seems to refer to EASA regulations	Registration available via e-citizen identification using email or using paper forms	Online registration and enrolling to exam using strong identification technology that usually requires Finnish bank account	No links, nor information through EASA site
Training	Online courses available in German language	Training instructions available in Bulgarian	Multiple links to different types of manuals, requirements, and guidance materials	Online training courses available in Finnish, Swedish, and English	Not available through EASA site
Applications for authorization	Application forms available, and to be emailed to Austro Control	Application forms available for some categories and for cross-border operations	List of forms and publications available in Croatian language	Application forms available in Finnish, Swedish, and English	Not available through EASA site

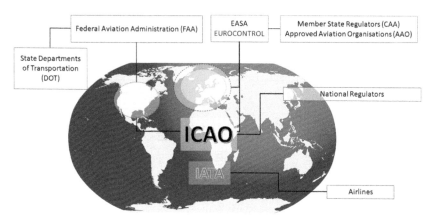

Figure 12.1 Global governance architecture for aviation—focus on Europe and the United States.

implementation and regular updating of the regulation framework. Also, insurance companies require empirical data and knowledge for liability issues. Insurance policies may dictate additional rules for operators and the role of insurance companies in defining limits of UAS operations may rise in the future (Stöcker et al., 2017).

The adoption of the new EU UAS regulations has itself included various issues, for example, multiple delays on both European and national levels. The two new regulations, Reg. (EU) 2019/947 and 2019/945, should have originally become applicable on July 1, 2020, but the date was delayed by the Commission to December 31, 2020 due to issues related to COVID-19 pandemic and consequent issues in implementation at member state level. The regulations also have a transition period for adopting the requirement for class identification label for UAS in open category. The duration of the transition period was originally set until December 31, 2022, but it was prolonged until December 31, 2023 due to lack of commercially available UAS with the class identification. On a national level, for example, Finnish competent authority postponed the open category A2 subcategory requirement for additional theoretical knowledge exam from original January 1, 2021 to April 1, 2021, and then to January 1, 2022.

Governance can also create critical bottlenecks for wider use of drones in various applications. For example, various applications such as drone photography and cinematography, surveying, and inspections can already be performed adequately in the current legislative environment that is largely promoting VLOS operations, however, they could likely benefit from BVLOS operations, whereas some applications such as search and rescue, UAS delivery service and efficient large scale agricultural use can be very much dependent on the ability to operate BVLOS. The European Parliament resolution on safe use of remotely piloted aircraft systems (2014/2243(INI)) from October 2015 already "stresses the importance of 'out-of-sight' flights for the development of sector" and "considers that European legislation should favour this modus operandi."

The public acceptance of UAS also shapes the governance in various ways. People working within the industry unlikely have hard time in distinguishing between commercial and military use-cases and operators, whereas within the public this is not guaranteed. As negative news tends to spread faster and wider than good ones, few bad actors can severely damage the public perception. Negative events may discourage investors or authorities from supporting the industry and move the governance toward unnecessarily tight restrictions. Poor public perception can also affect the enforcement of existing regulations, for example, the competent authority may not be that willing to green light a specific operation, if there is low public acceptance and there is a risk of sharing the blame if something goes wrong. It can also lower the willingness and demand to utilize UAS services. Thus improving and maintaining the public acceptance via, for example, transparency, education, community, and stakeholder outreach, should always be one priority of the industry.

Social acceptability as a term means a fairly broad set of factors such as risks, trust, values, benefits, expectations, knowledge, person background, impacts on environment, opportunity to take part in decision making, and so on. The acceptability of civil UAS among the majority of the population has been discussed, but there is still limited scientific research available. A commonly raised issue is the fear of a UAS falling to the ground and associated damage caused to people and property. In addition, people are afraid of losing their privacy. Most people are most concerned about monitoring activities around their home, then in the workplace, and less in public areas (Miethe et al., 2014). On a broader scale, acceptability plays a major role in considering increasing the potential for the use of drones. Some concerns exist that strong resistance could restrict or slow the wider dissemination of UAS applications. On the other hand, a positive atmosphere creates favorable conditions for increasing utilization. According to a study completed in 2020, people's attitudes were barely more positive (49%) than negative (43%), while rest (8%) were still "undecided" (Eißfeldt et al., 2020).

Since UASs have a wide variety of uses, social acceptability has been studied from different perspectives, like acceptance related to use of UAS in package delivery or how the population itself uses or how many owns UASs (e.g., Miethe et al., 2014; United States Postal Service, 2016). According to Miethe et al. (2014) and Eißfeldt et al. (2020), most people recognize UASs and know some uses for them, although there may be differences between countries. Most people associate the term "drone" with either military operations, surveillance and observation, videos and photos, leisure or hobby, accident or threat, or delivery and transport services. The general acceptability seems to be high for using UAS in rescue missions or in climate research and geographic mapping. The least support is granted for monitoring public places or package delivery services to private dwellings (Miethe et al., 2014; Eißfeldt et al., 2020). Researchers have found great variations in the acceptability of UAS depending on the age of the person, marital status, political orientation, income, and human rights attitudes. Among young and old people, UASs are more acceptable than among adults under 65 years of age, and men seems to have more positive attitude to drones than women (Eißfeldt et al., 2020). One way to increase acceptance is to increase people's knowledge and experience of UASs. However, studies show that this is not always enough, therefore legislation and regulations are of great importance.

Several threats and concerns related to the use of UAS has been identified by scholars throughout the past two decades. Some of the threats are still fictitious, which at some point can be realized and some are already real. Currently, it seems that most people feel the biggest risks coming from misuse of UAS or use in criminal activities. According to Eißfeldt et al. (2020), the following concerns are most typical in the order listed:

* crime and misuse,
* violation of privacy,
* liability and insurance,
* transport safety,
* damages and injuries,
* animal welfare,
* noise.

Researchers found that women and older respondents were more concerned than males and younger respondents. In addition, a correlation was found that people with more information about UAS were less concerned. Digitalization together with IoT (Internet of Things), AI (artificial intelligence), and other developed technologies raise threats in people's minds. In smart city, concepts with the use of drones several risks are identified (Rawat et al., 2022). Cyber and physical attacks concern the researchers most. Still, they believed that with the help of other technologies all the challenges can be solved.

Restrictive factors in agricultural applications of unmanned aircraft systems

Thinking of agricultural use cases of UAS, the differences in national implementation and administration, not to mention the obvious short-comings, will probably pose some thresholds in the specific and certified categories, making in some cases the market more domestic than EU-Single. For the open category, such challenges naturally do not exist if the operations can be carried out within the category requirements and in nonrestricted zones. The maximum take-off mass of 25 kg in the open category may, however, limit the ability to utilize larger UAS designed for spraying and dispersing applications. The ability of competent authority to designate separate UAS geographical zones, in which UAS operations are exempt from one or more of the requirements of open category, may alleviate the problem. However, this again is dependent on the national implementation on member state level.

UAS-based applications in agriculture largely revolve around either different remote sensing techniques or spraying and dispersing, both of which can face some of the same regulatory challenges but also some application-specific challenges. For example, regarding spraying applications it is not sufficient to only consider UAS regulation. For example, Directive 2009/128/EC prohibits aerial spraying of pesticides in EU member states with some derogations, for example, competent authority may permit aerial spraying if, for example, there is no technically feasible alternative method due to inaccessible terrain. Such derogations have been in use in few countries such as Germany and France where it has been possible to utilize aerial spraying, for example, in vineyards located in steep hillsides or riverbanks.

However, proposed new regulation (2022/0196/COD) would allow EU member state to exempt aerial application by an unmanned aircraft when factors related to the use of the unmanned aircraft demonstrate lower risks compared to risks arising from other aerial or land-based application equipment. These factors include, for example, the technical specifications of the unmanned aircraft, weather conditions, the area to be sprayed, and the level of training required for pilots, although precise criteria will be established when further evidence is available for science-based decision making.

Although 23% of Hungarian crop farms have reported using some precision farming techniques (close to EU average 25%), the use of UAS is still negligible. A study carried out by Bai et al. (2022) covered 200 Hungarian precision farmers, of which 17% utilized UASs. Compared to average farmer in Hungary, the farmers had larger farms (672 vs 20 ha), were slightly younger (53 vs 58 years), and had far higher education (49% had a university degree vs 3.4% on average). Interestingly, the share of females applying precision farming techniques in the sample was lower than the average among Hungarian farmers (6% vs 29%). Similar findings were made by Michels et al. (2021) in their sample of German farmers who were already aware of potential usage and applications of UAS in agriculture. The studies concluded that UAS were technologically mature enough for precision agriculture, but that mass application is slowed down due to challenges and questions related to legislation, investment, knowledge, and experience. Large farms are the most typical users of precision technologies and would also be the likely users of UAS technologies. These farms would probably benefit the most from more flexible regulatory approach to BVLOS operations.

The aforementioned low utilization of UAS in agriculture reflects most likely the broader situation in EU. Agriculture is slow in the adoption of new technologies (see, e.g., Leviäkangas and Kauppila, 2020), and UA technologies are probably no exception. However, the People's Republic of China has seen some rapid development in the utilization of agricultural UAS (Ipsos Business Consulting, 2019). As the country faces issues with aging and urbanizing population, and overuse of fertilizers and pesticides leading to soil degradation, UAS are seen as one part of modernizing the agriculture sector. First province scale subsidy programs for agriculture UAS began already in 2014 and first nationwide pilot subsidy scheme launched in 2017 (Ipsos Business Consulting, 2019). Also, between 2014 and 2017, the number of training centers increased 10-fold and granted UAS pilot certificates increased 100-fold. However, China's agriculture UAS industry has faced some of the same issues as their European counterpart, for example, regulatory restrictions limiting UAS operations to VLOS, lacking sufficient training to farmers to increase basic know-how, and lacking pilot training capacity to answer the demand.

Conclusions

By and large, it seems evident that even if the first global and supernational regulations have been set, there is much to be done at country level. National adoption of international regulations will for long be delayed in some countries, while in some other countries the progress could be faster but will be slowed down by the lack of international standards, agreements, and regulations.

One aspect that will be continuously and probably increasingly be of importance is safety and security. Drones have originally been developed for military purposes and human-induced risks will dominate the regulatory development. Yet, the commercial prospects and public good potential are substantial. This is one of the reasons why national regulations will be diversified in terms of substance, speed of adoption, and contents. There is no doubt either that some parts of UA regulations will have political loadings.

The technology push will be aggressive, and it is foreseeable that regulations and governance will not be able to keep up with the pace. Therefore the regulatory process will be reactive rather than proactive, and while the process attempts to be technology independent (as it should), it will need to react to different new technological applications as well as new use cases of UASs. The issue with reactive regulation is that it tends to provide neither flexibility nor stability. In sum, the governance strategy for UA will be trial-oriented (gain experiences), cautious (do not fix too many things), and reactive (see what happens).

One of the important issues for the governance and regulatory development is to collect data on incidents, accidents, and risky situations. Such reporting systems are not yet in place and need to be developed as knowledge and empirical data are accumulated. The reporting has to be mandated, operationalized, and quality controlled, which require setting up governance structures and processes—which will take time and effort. Yet, without reliable and working reporting systems it is hard to see that regulations can be well targeted and soundly argued.

Acknowledgments

This study was partly supported by the SPADE project, funded by the European Union's Horizon Europe Research and Innovation programme within HORIZON-CL6-2021-GOVERNANCE-01 under Grant Agreement no. 101060778, and K.H. Renlund Foundation.

References

Bai, A., Kovách, I., Czibere, I., Megyesi, B., Balogh, P., 2022. Examining the adoption of drones and categorisation of precision elements among hungarian precision farmers using a trans-theoretical model. Drones 6. Available from: https://doi.org/10.3390/drones6080200.

Bassi, E., 2020. From here to 2023: civil drones operations and the setting of new legal rules for the European single sky. J. Intell. Robot. Syst. Theory Appl. 100, 493−503. Available from: https://doi.org/10.1007/s10846-020-01185-1.

DroneDeploy, 2022. DroneDeploy's State of the Drone Industry Report 2022. Available online: https://www.dronedeploy.com/resources/ebooks/state-of-the-drone-industry-report-2022/ (accessed on 16 December 2022).

Du, H., Heldeweg, M.A., 2019. An experimental approach to regulating non-military unmanned aircraft systems. Int. Rev. Law Comput. Technol. 33, 285−308. Available from: https://doi.org/10.1080/13600869.2018.1429721.

Eißfeldt, H., Vogelpohl, V., Stolz, M., Papenfuß, A., Biella, M., Belz, J., et al., 2020. The acceptance of civil drones in Germany. CEAS Aeronaut. J. 11, 665−676. Available from: https://doi.org/10.1007/s13272-020-00447-w.

Fortune Business Insights, 2020. Agriculture Drones Market. Available online: https://www.fortunebusinessinsights.com/agriculture-drones-market-102589 (accessed on 16 December 2022).

Hernández, G., Amaral, M., 2022. Case studies on agile regulatory governance to harness innovation: civilian drones and bio-solutions. OECD Regulatory Policy Working Papers, No. 18, OECD Publishing, Paris. Available from: https://doi.org/10.1787/0fa5e0e6-en.

Ipsos Business Consulting, 2019. China's Agriculture Drone Revolution: Disruption in the Agriculture Ecosystem. Available online: https://www.ipsos.com/sites/default/files/ct/publication/documents/2020-10/china-agriculture-drones.pdf (accessed 16 December 2022).

JARUS, 2019. JARUS Guidelines on Specific Operations Risk Assessment (SORA), 2.0 ed. Available online: http://jarus-rpas.org/content/jar-doc-06-sora-package (accessed on 16 December 2022).

Leviäkangas, P., Kauppila, O., 2020. Digitalisation of industries: a comparative analysis from Australia and Finland. Int. J. Technol. Policy Manag. 20, 70−89. Available from: https://doi.org/10.1504/IJTPM.2020.104868.

Michels, M., von Hobe, C.F., Weller von Ahlefeld, P.J., Musshoff, O., 2021. The adoption of drones in German agriculture: a structural equation model. Precis. Agric. 22, 1728−1748. Available from: https://doi.org/10.1007/s11119-021-09809-8.

Miethe, T.D., Lieberman, J.D., Sakiyama, M., Troshynski, E.I., 2014. Public Attitudes about Aerial Drone Activities: Results of a National Survey. State Data Brief. Center for Crime and Justice Policy: Las Vegas, NV. CCJP 2014-02.

Nex, F., Armenakis, C., Cramer, M., Cucci, D.A., Gerke, M., Honkavaara, E., et al., 2022. UAV in the advent of the twenties: where we stand and what is next. ISPRS J. Photogramm. Remote Sens. 184, 215−242. Available from: https://doi.org/10.1016/j.isprsjprs.2021.12.006.

Öz, E., Heikkilä, E., Tiusanen, R., 2022. Development of an organisational certification process for specific category drone operations. Drones 6. Available from: https://doi.org/10.3390/drones6100278.

Pagallo, U., Bassi, E., 2020. The governance of unmanned aircraft systems (UAS): aviation law, human rights, and the free movement of data in the EU. Minds Mach. 30, 439−455. Available from: https://doi.org/10.1007/s11023-020-09541-8.

Presedence Research, 2022. Commercial Drone Market. Available online: https://www.precedenceresearch.com/commercial-drone-market (accessed on 16 December 2022).

Rawat, B., Bist, A.S., Apriani, D., Permadi, N.I., Nabila, E.A., 2022. AI based drones for security concerns in smart cities. APTISI Trans. Manag. 7, 125−130. Available from: https://doi.org/10.33050/atm.v7i2.1834.

Researchdive, 2022. UAV Drones Market Report. Available online: https://www. researchdive.com/8348/unmanned-aerial-vehicle-uav-drones-market (accessed on 16 December 2022).

Stöcker, C., Bennett, R., Nex, F., Gerke, M., Zevenbergen, J., 2017. Review of the current state of UAV regulations. Remote Sens. 9, 33−35. Available from: https://doi. org/10.3390/rs9050459.

Tsiamis, N., Efthymiou, L., Tsagarakis, K.P., 2019. A comparative analysis of the legislation evolution for drone use in OECD countries. Drones 3, 1−15. Available from: https://doi.org/10.3390/drones3040075.

United States Postal Service, 2016. Public Perception of drone delivery in the United States. RARC Rep. 114.

Index

Printed and bound by CPI Group (UK) Ltd, Croydon, CR0 4YY

03/10/2024

01040847-0006